# Micro-Optics and Energy

Jacob J. Lamb • Bruno G. Pollet
Editors

# Micro-Optics and Energy

## Sensors for Energy Devices

Springer

*Editors*
Jacob J. Lamb
Department of Electronic Systems, ENERSENSE
Norwegian University of Science and Technology
Trondheim, Norway

Department of Energy and Process Engineering, ENERSENSE
Norwegian University of Science and Technology
Trondheim, Norway

Bruno G. Pollet
Department of Energy and Process Engineering, ENERSENSE and NTNU Team Hydrogen
Norwegian University of Science and Technology
Trondheim, Norway

ISBN 978-3-030-43675-9 ISBN 978-3-030-43676-6 (eBook)
https://doi.org/10.1007/978-3-030-43676-6



This Springer imprint is published by the registered company Springer Nature Switzerland AG.
The registered company address is: Gewerbestrasse 11, 6330 Cham, Switzerland

# Preface

More than 80% of our present energy consumption is chemical and non-renewable (coal, oil and gas). The most important renewable energy alternatives (wind and solar energy) produce intermittent electric energy and are also inadequate for most transportation options. The renewable transition will require a collective effort using many different types of energy conversion and storage devices and technologies to entirely remove the dependence on non-renewable fossil fuels. Many technologies have only recently entered the commercial market, with many others still not yet commercialised. This is due to the requirement for technologies to be researched further to understand how they can be improved. Sensors for the measurement and monitoring of energy conversion and storage devices are needed to improve our understanding of such technologies.

This volume intends to provide a brief research source for micro-optical sensors and energy conversion and storage devices, discussing fundamental aspects as well as cutting-edge trends. This volume provides industry professionals, researchers and students with the most updated review on modern energy conversion and storage technologies, as well as micro-optical sensors. This volume aims to help readers identify technology gaps and develop new materials and novel designs that lead to commercially viable non-fossil energy systems.

The editors and authors are grateful to the ENERSENSE programme, the ENERSENSE team and NTNU Team Hydrogen at the Norwegian University of Science and Technology (NTNU) for supporting and helping on this book volume.

Trondheim, Norway

Jacob J. Lamb
Bruno G. Pollet

# Acknowledgements

The authors are grateful to the ENERSENSE programme and NTNU Team Hydrogen at the Norwegian University of Science and Technology (NTNU) for supporting and helping on this book project.

# Contents

# Editors and Contributors

## About the Editors

**Jacob J. Lamb** is originally from New Zealand where he got both his B.Sc. and M.Sc. in Biochemistry at the University of Otago, where he worked in a research laboratory at the University of Otago in Dunedin, New Zealand with Professor Julian Eaton-Rye. He came to Norway to undertake a Ph.D. in Biotechnology under the supervision of Associate Professor Martin Hohmann-Marriott, which he completed in June 2016. Jacob then undertook a post-doctoral position under the supervision of Professor Dag Roar Hjelme and Associate Professor Kristian Myklebust Lien looking at photonic sensor technologies for their application in biogas production. Since 2018, Jacob has been a researcher studying sensor technologies and process digitalisation for a variety of applications in biological and engineering fields related to energy. His areas of expertise include sensor technologies, optical spectroscopy, photosynthesis, microbiology, biological and biochemical techniques, electronics and programming, digitalisation, renewable energy, energy storage and process engineering. His research motivation is to improve renewable energy sources, increase sustainability within agricultural and aquacultural industries, develop technologies for climate change mitigation as well as develop ways to measure, analyse and optimise processes. He has also worked on implementing a knowledge base of Li-ion battery manufacturing at NTNU.

**Bruno G. Pollet** is a full Professor of Renewable Energy at the Department of Energy and Process Engineering at the Norwegian University of Science and Technology (NTNU) in Trondheim. His research covers a wide range of areas in electrochemical engineering, electrochemical energy and sono-electrochemistry (the use of power ultrasound in electrochemistry) from the development of novel materials and hydrogen fuel cell to water treatment/disinfection demonstrators and prototypes. He was a Professor of Energy Materials and Systems at the University of the Western Cape (South Africa) and R&D Director of the National Hydrogen South Africa (HySA) Systems Competence Centre. He was also a co-founder and an Associate Director of the University of Birmingham Centre for Hydrogen and Fuel Cell Research in the UK. He was awarded a Diploma in Chemistry and Material

Sciences from the Université Joseph Fourier (France), a B.Sc. (Hons) in Applied Chemistry from Coventry University (UK) and an M.Sc. in Analytical Chemistry from the University of Aberdeen (UK). He also gained his Ph.D. in Physical Chemistry in the field of electrochemistry and sonochemistry at the Coventry University Sonochemistry Centre.

## Contributors

**Robert Bock** Faculty of Engineering, Department of Energy and Process Engineering, ENERSENSE, Norwegian University of Science and Technology, Trondheim, Norway

**Odne S. Burheim** Department of Energy and Process Engineering, ENERSENSE and NTNU Team Hydrogen, Norwegian University of Science and Technology, Trondheim, Norway

**Eline F. J. Gregoire** Faculty of Engineering, Department of Energy and Process Engineering, Norwegian University of Science and Technology, Trondheim, Norway

**Truls Gundersen** Faculty of Engineering, Department of Energy and Process Engineering, Norwegian University of Science and Technology, Trondheim, Norway

**Astrid S. Gunnarshaug** Faculty of Natural Sciences, Department of Chemistry, Norwegian University of Science and Technology, Trondheim, Norway

**Dag R. Hjelme** Department of Electronic Systems, ENERSENSE, Norwegian University of Science and Technology, Trondheim, Norway

**Md H. Islam** Faculty of Engineering, Department of Energy and Process Engineering, ENERSENSE, Norwegian University of Science and Technology, Trondheim, Norway

**Kjersti W. Krakhella** Faculty of Natural Sciences, Department of Materials Science and Engineering, ENERSENSE, Norwegian University of Science and Technology, Trondheim, Norway

**Jacob J. Lamb** Department of Electronic Systems, ENERSENSE, Norwegian University of Science and Technology, Trondheim, Norway

Department of Energy and Process Engineering, ENERSENSE, Norwegian University of Science and Technology, Trondheim, Norway

**Kristian M. Lien** Faculty of Engineering, Department of Energy and Process Engineering, ENERSENSE, Norwegian University of Science and Technology, Trondheim, Norway

**Zhongxuan Liu** Faculty of Engineering, Department of Energy and Process Engineering, Norwegian University of Science and Technology, Trondheim, Norway

Faculty of Engineering, Department of Language and Literature, Norwegian University of Science and Technology, Trondheim, Norway

**Harald I. Muri** Faculty of Information Technology and Electrical Engineering, Department of Electronic Systems, ENERSENSE, Norwegian University of Science and Technology, Trondheim, Norway

**Eilif S. Øyre** Faculty of Engineering, Department of Energy and Process Engineering, ENERSENSE, Norwegian University of Science and Technology, Trondheim, Norway

**Bruno G. Pollet** Department of Energy and Process Engineering, ENERSENSE and NTNU Team Hydrogen, Norwegian University of Science and Technology, Trondheim, Norway

**Yash D. Raka** Faculty of Engineering, Department of Energy and Process Engineering, ENERSENSE, Norwegian University of Science and Technology, Trondheim, Norway

**Rolf K. Snilsberg** Faculty of Information Technology and Electrical Engineering, Department of Electronic Systems, ENERSENSE, Norwegian University of Science and Technology, Trondheim, Norway

**Lena Spitthoff** Faculty of Engineering, Department of Energy and Process Engineering, ENERSENSE, Norwegian University of Science and Technology, Trondheim, Norway

**Federico Ustolin** Faculty of Engineering, Department of Mechanical and Industrial Engineering, Norwegian University of Science and Technology, Trondheim, Norway

**Markus S. Wahl** Faculty of Information Technology and Electrical Engineering, Department of Electronic Systems, ENERSENSE, Norwegian University of Science and Technology, Trondheim, Norway

# Part I

# Optical Properties for Sensors

# Introduction

# 1

Jacob J. Lamb, Odne S. Burheim, and Dag R. Hjelme

In many recent research projects, research is conducted between several different fields, with experts working together to achieve a common goal. The development of nanostructures for use in battery technologies is one example, where the fields of electrochemistry and nanoscience have come together to achieve a common goal. In order to understand and develop research that integrates multiple fields of research, experts in the required fields must come together. Concerning modelling and simulations, the computation requires variable verification through experiments. This relies on measuring the most critical properties in sufficient detail in terms of numerical and geometrical precision. Electrochemical energy conversion and storage devices are one such case where measurements of high resolution, in regard to their geometrical precision within a device, are required.

Optical fibre-based sensors have the potential for being used on a microscopic geometric scale with very high precision. With electrochemical energy storage devices becoming thinner in order to improve their performance, the size requirements for sensor systems is in the range of 5–500 μm. This book is intended to uncover the possibilities and requirements to integrate the fields of optical fibre-

J. J. Lamb (✉)
Department of Electronic Systems, ENERSENSE, Norwegian University of Science and Technology, Trondheim, Norway

Department of Energy and Process Engineering, ENERSENSE, Norwegian University of Science and Technology, Trondheim, Norway
e-mail: jacob.j.lamb@ntnu.no

O. S. Burheim
Department of Energy and Process Engineering, ENERSENSE and NTNU Team Hydrogen, Norwegian University of Science and Technology, Trondheim, Norway
e-mail: odne.s.burheim@ntnu.no

D. R. Hjelme
Department of Electronic Systems, ENERSENSE, Norwegian University of Science and Technology, Trondheim, Norway
e-mail: dag.hjelme@ntnu.no



J. J. Lamb, B. G. Pollet (eds.), *Micro-Optics and Energy*,
https://doi.org/10.1007/978-3-030-43676-6_1

based sensors in electrochemical energy storage systems, allowing the improvement of variable resolution in their experimental model verification research.

## Dimensions of Electrochemical Energy Storage Devices

Energy is a property defined by mass, geometry and time. The units of energy (e.g. Joule, Nm or kg $m^2s^{-2}$) illustrate this. Energy is exchanged by means of heat and work, as defined by the laws of thermodynamics. It can be stored or change form. In electrochemical (and electric) energy storage, electrical work is converted into latent work (a chemical potential) and heat. Heat, beyond reversible heat ($T\Delta S$), is an indication of lost work from various forms of friction during energy conversion or transport processes. This heat (or loss of work) increases with the rate of energy conversion. Energy storage devices convert energy in order to store or release energy, and modern energy storage devices are often developed to convert energy more effectively (faster), rather than with higher efficiency [1].

Energy conversion between electric energy and chemical energy relies on electrochemical cells, such as lithium-ion batteries, supercapacitors, water electrolysis and hydrogen fuel cells [2]. These are reactors where thin layers of active and passive components are assembled in several layers. Typically, the active layers have a thickness below 200 μm, with many as thin as 10–20 μm [3–6].

Inside these rather thin layers, the conversion of electrical and chemical energy takes place. Increases in temperature due to irreversible reactions or friction within these materials that generally have a rather low thermal conductivity (0.05–1 W $K^{-1}$ $m^{-1}$). In terms of chemical processes, degradation mechanisms begin to occur and increase as the temperature increases [7, 8]. These reactions can be traced or sensed by highly localised temperature measurements. Understanding such processes to a high resolution within such small geometric spaces requires applying sufficiently small sensors that are at the same time inert to the ongoing processes. In several cases, optical fibres can be tuned concerning size and detection variable without affecting the quality of the sensor. It is fair to say that optical fibre-based sensors are a developing tool for improved understanding and model verification within electrochemical systems. An overview of the electrochemical technologies and possible optical fibre sizes is shown in Fig. 1.1.

## Electrical Versus Optical Sensors

The detection of external stimuli on electronic sensors can often be segregated into the terms active or passive sensing. By the use of an excitation signal, the active sensor will change in response to an external effect and produce the output in the form of the change in current or voltage. A thermistor is a typical example of an active temperature sensor. For passive sensing, there is no need for an excitation signal to produce the current or voltage in response to the external stimuli. Thermocouples are often used as passive temperature sensors. Sensors based on

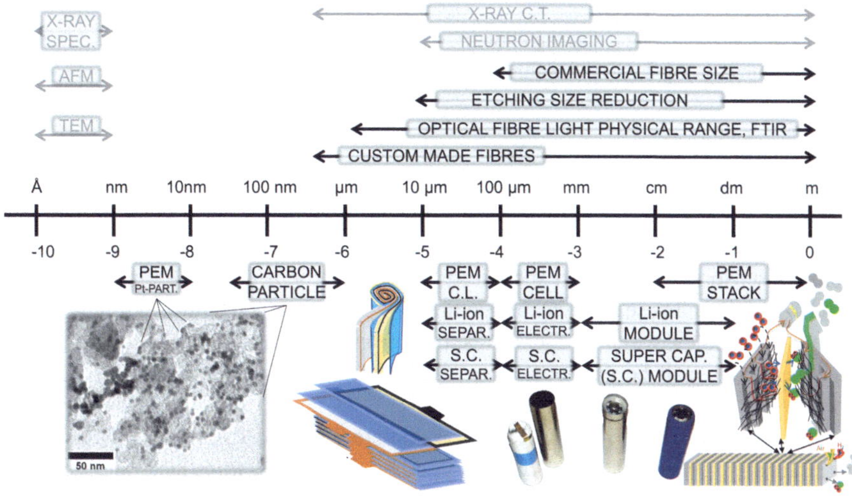


**Fig. 1.1** Overview of different electrochemical devices and possible characterisation technologies with emphasis on optical fibres

electronic circuit systems are convenient, cheap and simplified but have limitations when used for monitoring harsh environments such as oxidising (pH too) fluids, in high temperatures or high pressure, and have a size demand.

Optical fibre (OF) sensors for temperature, pressure and strain sensing have been utilised in many industrial applications in the last decade. Combined with the robustness of the sensor design for monitoring in harsh environments, OF sensors are also immune to electromagnetic interference. Active or passive sensing with OF sensors offers remote multiplexed and multipoint sensing capabilities, simplified design and real-time monitoring of temperature, strain, humidity or concentration of a specific chemical in complex mixtures. Active sensing with OFs utilises a light source such as a laser or a broadband source as the excitation signal that responds to an external stimulus measured by a photodetector. In passive sensor with OFs, the excitation signal or light source is omitted so that only the photodetector detects the background light. Active or passive OFs are organised in intrinsic or extrinsic sensor configurations. Extrinsic OF sensors monitor the medium in the exterior of the OF; whereas the intrinsic OF sensors monitor the interior medium that responds to changes in the exterior medium [9].

## General Principles of Optical Fibre Sensor Systems

A sensor is often understood as the transducer alone, the part of the sensor that converts the measured quantity into an electrical signal. In an optical sensor, the transducer creates an optical signal, which then also requires an optical-to-electrical conversion. Figure 1.2 shows an illustration of the complete sensor system.

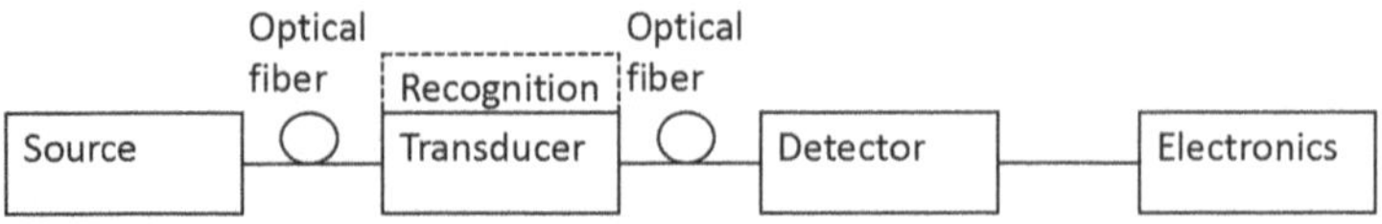


**Fig. 1.2** General components needed for a sensor setup

Although the sensor sensitivity can be defined based on the transducer alone, the resolution and accuracy of the sensor are often dependent on the electronics. For chemical sensors, the required specificity introduces the need for recognition of the correct variable. These analyte molecules typically interact with binding sites or receptor molecules, which triggers a response from the transducer. Together with the transducer, this will be the focus of a significant part of this book, which will be limited to fibre optic sensors.

The transducer shown in Fig. 1.2 can modulate the optical signal in various ways (e.g. intensity, phase and polarisation), based on the source, transducer and detector type used. A common way of doing this is through refractive index (RI) sensing, where the recognition element converts the analyte response into RI changes that are detected by the transducer.

The recognition is typically realised with a stimuli-responsive polymer [10] or surface plasmon-generating layer of metal [11], and typical RI transducers include long-period gratings (LPGs) [12], Mach–Zehnder type interferometers [13] and tilted fibre Bragg gratings (TFBG) [14–16]. These sensors have factor that they excite cladding modes, of which their phase is sensitive to the surrounding RI, in common [17, 18]. The changed phase will result in an interference signal when recombined with the core mode. Sensing of pH can be realised by coating an RI-sensitive sensor with a polymer that changes its optical density based on pH [13, 18]. Linear response both in acidic and alkaline solutions has been achieved using these techniques [13]. Although these sensors are temperature dependent, the sensitivity is an order of magnitude lower than that in LPGs.

## Sensor Integration

Interrogation methods and components applied in OF sensor systems are often specially designed for the parameters used for the monitoring. To make the sensor system compatible with existing regulation techniques in fuel cells and electrolysers, the components used for signal excitation, modulation and acquisition should be designed without disrupting and reducing the efficiency of the energy storage system. For an OF embedded in a cathode catalyst layer of a proton exchange membrane fuel cell (PEMFC) for multipoint temperature measurements, the fibre itself will disrupt the thickness of the layer by occupying a space equal to 125 μm in diameter of a cylinder. With a large surface area of the layers of a PEMFC, an OF embedded in the layer will result in small disruption of the gas diffusion or the transportation of heat.

By monitoring the temperature along the OF (i.e. along a one-dimensional line in the PEMFC), the regulation of the temperature can be controlled by cooling channels along the same line. Therefore, for OFs in a mesh, the temperature of the cooling channels can be controlled along with the same mesh. The inputs and outputs of the signal have to be centralised by a control unit depending on if it is a reflection- or transmission-based OF sensor system. Sensor fusion can be applied in such a mesh system based on a model calibrated for that particular fuel cell in response to controlled temperature changes.

This book intends to give an overview of selected energy conversion and storage devices as well as describing essential optical properties required to be considered when coupling OF-based sensors for parameter detection. Examples of integration of OF sensors into energy conversion and storage devices are also detailed to display the ability afforded with such sensor networks.

**Acknowledgements** The authors are grateful to the ENERSENSE programme and NTNU Team Hydrogen at the Norwegian University of Science and Technology (NTNU) for supporting and helping on this book project.

## References

1. Burheim OS, Pharoah JG (2017) A review of the curious case of heat transport in polymer electrolyte fuel cells and the need for more characterisation. Curr Opin Electrochem 5(1):36–42
2. Burheim OS (2017) Hydrogen for energy storage. In: Engineering energy storage. Academic, Amsterdam, pp 147–192
3. Hauge HH, Presser V, Burheim O (2014) In-situ and ex-situ measurements of thermal conductivity of supercapacitors. Energy 78:373–383
4. Richter F, Kjelstrup S, Vie PJS, Burheim OS (2017) Thermal conductivity and internal temperature profiles of Li-ion secondary batteries. J Power Sources 359:592–600
5. Burheim OS (2017) Review: PEMFC materials' thermal conductivity and influence on internal temperature profiles. ECS Trans 80(8):509–525
6. Marshall AT, Sunde S, Tsypkin M, Tunold R (2007) Performance of a PEM water electrolysis cell using IrxRuyTazO2 electrocatalysts for the oxygen evolution electrode. Int J Hydrog Energy 32(13):2320–2324
7. Burheim OS, Onsrud MA, Pharoah JG, Vullum-Bruer F, Vie PJS (2014) Thermal conductivity, heat sources and temperature profiles of Li-ion batteries. ECS Trans 58(48):145–171
8. Burheim OS, Aslan M, Atchison JS, Presser V (2014) Thermal conductivity and temperature profiles in carbon electrodes for supercapacitors. J Power Sources 246:160–166
9. Lamb JJ, Bernard O, Sarker S, Lien KM, Hjelme DR (2019) Perspectives of surface plasmon resonance sensors for optimized biogas methanation. Eng Life Sci 19:759–769
10. Tierney S, Volden S, Stokke BT (2009) Glucose sensors based on a responsive gel incorporated as a Fabry-Perot cavity on a fiber-optic readout platform. Biosens Bioelectron 24(7):2034–2039
11. Renoirt J-M, Zhang C, Debliquy M, Olivier M-G, Mégret P, Caucheteur C (2013) High-refractive-index transparent coatings enhance the optical fiber cladding modes refractometric sensitivity. Opt Express 21(23):29073–29082
12. James SW, Tatam RP (2003) Optical fibre long-period grating sensors: characteristics and application. Meas Sci Technol 14(5):R49
13. Gu B, Yin M-J, Zhang AP, Qian J-W, He S (2009) Low-cost high-performance fiber-optic pH sensor based on thin-core fiber modal interferometer. Opt Express 17(25):22296–22302

14. Chen C, Caucheteur C, Mégret P, Albert J (2007) The sensitivity characteristics of tilted fibre Bragg grating sensors with different cladding thicknesses. Meas Sci Technol 18(10):3117
15. Chan C-F, Chen C, Jafari A, Laronche A, Thomson DJ, Albert J (2007) Optical fiber refractometer using narrowband cladding-mode resonance shifts. Appl Opt 46(7):1142–1149
16. Laffont G, Ferdinand P (2001) Tilted short-period fibre-Bragg-grating-induced coupling to cladding modes for accurate refractometry. Meas Sci Technol 12(7):765
17. Li E, Wang X, Zhang C (2006) Fiber-optic temperature sensor based on interference of selective higher-order modes. Appl Phys Lett 89(9):91119
18. Wong WC, Chan CC, Hu P, Chan JR, Low YT, Dong X et al (2014) Miniature pH optical fiber sensor based on waist-enlarged bitaper and mode excitation. Sensor Actuat B-Chem 191:579–585

# Light Properties and Sensors

# 2

Markus S. Wahl, Jacob J. Lamb, Harald I. Muri, Rolf K. Snilsberg, and Dag R. Hjelme

## Light as Electromagnetic Waves

Light has the properties of both particles and waves, depending on the situation. In Young's double-slit interferometer, light behaves like waves, creating a pattern similar to what one could make with water ripples in a pond.

On the other hand, the photoelectric effect proves that light consists of particles—each with an energy of $E = hv$, where $h$ is Planck's constant and $v$ is the light frequency. The energy of light with the specific frequency $v$ is quantised as a photon that represents the "particle." The wave–particle duality can be observed in the Youngs double slit and the photoelectric effect.

Light as a wave contains frequency, amplitude and phase; whereas, light as a photon contains quantised energies and different momentum for different frequencies. Light can also be described in the so-called Ray model, where only simple geometrical laws are used to determine the refraction of light in an optical imaging system. Further in this chapter, we will talk about the light as electromagnetic (EM) radiation, where the EM term describes the light containing an electric and a magnetic component in a propagating wave.

The EM wave creates charges and currents when propagating in a dielectric or metal media (e.g. glass or gold), and carries energy and momentum that is dependent

M. S. Wahl · H. I. Muri · R. K. Snilsberg · D. R. Hjelme
Department of Electronic Systems, ENERSENSE, Norwegian University of Science and Technology, Trondheim, Norway
e-mail: markus.s.wahl@ntnu.no; harald.muri@ntnu.no; rolf.k.snilsberg@ntnu.no; dag.hjelme@ntnu.no

J. J. Lamb (✉)
Department of Electronic Systems, ENERSENSE, Norwegian University of Science and Technology, Trondheim, Norway

Department of Energy and Process Engineering, ENERSENSE, Norwegian University of Science and Technology, Trondheim, Norway
e-mail: jacob.j.lamb@ntnu.no



J. J. Lamb, B. G. Pollet (eds.), *Micro-Optics and Energy*,
https://doi.org/10.1007/978-3-030-43676-6_2

**Fig. 2.1** Electromagnetic spectrum

on the wavelength. The light interacting with charges and currents in materials are also often referred to as EM optics. In EM optics, Maxwell's equations are used to solve the electric and magnetic field vectors for position and time, where the material or the medium properties serve as the input values. The light–matter interactions in this model are explained as an EM wave interacting with charges and currents in the conducting materials.

The electromagnetic spectrum ranges from wavelengths 0.003 nm (gamma rays) to $10^8$ m (long radio waves). The wavelength range used for optical sensors can be divided into categories of ultraviolet, visible light that our eye can see, near-infrared (NIR) and mid-infrared (MIR) (Fig. 2.1).

The short and long light wavelengths may interact with materials in different ways. In the MIR or NIR range, the light may interact with the vibrational states of molecules; whereas, in the UV or VIS range the light may interact with the electron clouds of atoms or molecules. For some wavelengths (e.g. in the NIR range), the light can be confined and guided over long distances with small losses in silica-based waveguides. To explain other effects such as interference or diffraction, we can have a look at the electric component of the EM wave expressed as a function of time $t$ and space $\vec{z}$:

$$\vec{E}(\vec{z},t) = \mathrm{Re}\left\{\vec{A}\ \exp\left[j(k\vec{z} - \omega t)\right]\right\} = \vec{a}\ \cos\ \left(k\vec{z} - (\omega t + \varphi)\right) \tag{2.1}$$

Here $\vec{A} = \vec{a}\ \exp\ (j\varphi)$ is the complex envelope, $\omega$ is the angular frequency of the light, $k = 2\pi/\lambda$ is the wavenumber and $\varphi$ is the phase. The electric wave may be understood well by observing it for $t$ and $k$ individually. With $t = 0$ an immobilised electric cosine wave is observed and distributed along $\vec{z}$ with a period determined by the propagation constant $k$. When the value $k\Delta\vec{z} = (2\pi/\lambda)\,\Delta\vec{z} = 2\pi$, then $\Delta\vec{z}$ represents the distance equal to the wavelength period. The value of $1/k$ can then be understood as the distance in radians equal to the period of the wave. With $\vec{z} = 0$, we are observing the cosine wave for a fixed point in space with a frequency determined by $\omega$. With $\omega\Delta t = 2\pi v\Delta t = 2\pi$, then $\Delta t$ represents the time it takes for the wave to complete one cycle. Therefore, $\omega$ represents the frequency of the cosine-wave in radians.

Given the EM wave, we can now understand the effect of interference by first noting the phase travelled by the waves with the same frequency as $\varphi = kz$. One wave with phase $\varphi_1 = kz_1$ may have constructive interference with another wave with the same frequency but different phase (e.g. $\varphi_2 = kz_2$), so that the difference is $\varphi_2 - \varphi_1 = q2\pi$, where $q = 0,\ 1,\ 2$ are integers. For destructive interference, the difference in phase needs to be half the period of the wave so that $\varphi_2 - \varphi_1 = q\pi$.

The light diffraction effect can be explained given the interference of waves with the same wavelength but a different phase. When EM waves enter a slit or an obstacle with normal incidence, the waves are refracted for many angular degrees $\theta$. Here, the phase for each wave can be represented by phase $\varphi = k \sin \theta z$, where $\theta$ ranges from 0 to 90 degrees. The difference in propagation lengths $z$ and in angular degrees $\theta$ creates phase differences $\varphi_2 - \varphi_1$, with $q2\pi$ as constructive interference and $q\pi$ as destructive interference. For EM waves with different wavelengths, there are now three variables changing the phase $\varphi = k \sin \theta z = (2\pi/\lambda) \sin \theta z$, and the interference will not only vary with the $\theta$ and $z$, but also with the colour of the light. One example of polychromatic light diffraction is the reflection of light by a CD or DVD that appears as a rainbow to the eye.

Despite this, light refraction by a prism is due to another effect that is different from diffraction. Each EM wave with different wavelengths will propagate in the prism with a different RI so that phase $\varphi_1 = k_1 \sin \theta_1 n_1 z = (2\pi/\lambda) \sin \theta_1 n_1 z$, is different to $\varphi_2 = k_2 \sin \theta_2 n_2 z = (2\pi/\lambda) \sin \theta_2 n_2 z$. When white light is incident on the prism, the light will refract with different angles for different wavelengths. A wave with a normal incidence on the prism will have the phase at the air-glass interface as $\varphi_{air} = \varphi_{prism}$, so that $(2\pi/\lambda) \sin 90\, n_{air}\, z = (2\pi/\lambda) \sin \theta_{refract}\, n\, (\lambda)\, z$. It can be observed from $\theta_{refract} = n_{air}/n\, (\lambda)$ that the refraction angle $\theta_{refract}$ will with be dependent on wavelength due to $n\, (\lambda)$, which will change with different wavelengths.

Absorption, scattering, refraction or interference of light can be measured by detecting its intensity or energy for one or several wavelengths. There are several definitions of intensity or energy of light. Radiant flux expresses the energy emitted per unit time (W), and spectral flux expresses radiant flux per frequency or wavelength (W/Hz). To include the angle of incidence of light we can express radiant or spectral intensity, which is flux per steradian (W/sr) or flux per steradian and wavelength (W/(sr Hz)). Lastly, we have radiance or spectral radiance that takes into account the radiant or spectral intensity reflected, transmitted or received by a surface area of $m^2$ ($W/(sr\ m^2)$) and ($W/(sr\ Hz\ m^2)$), respectively.

## Mathematical Formalism

The following section will discuss the mathematical formalism of EM fields. The main areas of focus will include field strength and intensity, wave interference and polarisation.

### Field Strength and Intensity

As described in Eq. (2.2), the electric field of a light wave can be expressed as:

$$\vec{E}(\vec{z}, t) = \mathrm{Re}\left\{\overrightarrow{E_m}(z) \exp(i\omega t)\right\} \tag{2.2}$$

with a complex amplitude $\overrightarrow{E_m} = \vec{a}\ \exp\left[i\left(\vec{k}z + \varphi\right)\right]$. The magnetic field $\vec{H}$ can be expressed in a similar way. The direction of energy flux is given by the Poynting vector that is defined as:

$$\vec{S} = \vec{E} \times \vec{H} \tag{2.3}$$

which denotes the energy per unit area per unit time. For a sinusoidal plane wave, the electric and magnetic field oscillate in phase, where the Poynting vector always points in the direction of propagation with an oscillating magnitude. The intensity, described by the time-averaged magnitude of the electromagnetic field, can then be expressed as:

$$\langle S \rangle = \frac{1}{2\eta} |E_m|^2 \tag{2.4}$$

where $\eta$ is the electromagnetic impedance of the medium and $E_m$ is the complex amplitude. Therefore, information about the phase is not available in the measured intensity. By interference, with a second identical wave, the phase can be recovered.

### Interference of Monochromatic Waves

Interference between two monochromatic waves can be expressed as the time-averaged sum of the two complex amplitudes:

$$I_{\text{tot}} = \left\langle \left| \vec{E}_1 + \vec{E}_2 \right| \right\rangle^2 = \langle |E_1^2| \rangle + \langle |E_2^2| \rangle + \langle E_1 E_2^* \rangle + \langle E_2 E_1^* \rangle \tag{2.5}$$

The following equation (Eq. 2.6) can be applied when there are two monochromatic waves with the same polarisation, and can be written as:

$$I_{\text{tot}} = I_1 + I_2 + \sqrt[2]{I_1 I_2} \cos(\Delta\varphi) \tag{2.6}$$

where $\Delta\varphi$ is the phase difference between the two waves.

### Polarisation

Electromagnetic plane wave in the $z$-direction can be expressed as a sum of two waves polarised in the $x$- and $y$-direction:

$$\vec{E}(\vec{z}, t) = E_x \vec{x} + E_y \vec{y} \tag{2.7}$$

where

$$E_x = a_x \left[\omega \left(t - \frac{z}{c}\right) + \varphi_x\right] \tag{2.8}$$

$$E_x = a_y \cos\left[\omega\left(t - \frac{z}{c}\right) + \varphi_x\right] \tag{2.9}$$

Squaring and adding these values yields the equation for an ellipse:

$$\left(\frac{E_x}{a_1}\right)^2 + \left(\frac{E_y}{a_2}\right)^2 - 2\left(\frac{E_x}{a_1}\right)\left(\frac{E_y}{a_2}\right)\cos\varphi = \sin^2\varphi \tag{2.10}$$

with $\varphi = \varphi_y - \varphi_x$, given that $E_x = E_y$, the equation can describe a line, a circle or an ellipse, depending on the phase difference. Materials that delay the phases of the orthogonal modes differently are called birefringent and can be used to control the light polarisation.

Light polarisation is usually defined according to the plane of incidence. p-polarised light has its electric field parallel to this defined plane, while for s-polarised light it is perpendicular to this plane. Alternatively, these waves are denoted as transversal magnetic (TM) and transversal electric (TE), respectively. Reflection and transmission are typically polarisation dependent, but if the angle of incidence is zero, the two polarisations become degenerate, and reflection and transmission will be equal.

## Interaction of Light with Materials

Interaction of light with materials covers the generation of light, propagation (and absorption) of light through media and the detection of light. These three processes form the basis for all-optical sensing. Generation and detection of light are typically done with semiconductor materials, while propagation is typically described in a dielectric (e.g. glass).

### Dielectric Materials

Dielectric materials are insulators that can be polarised with an applied electric field. The electric field causes the bound charge carriers to shift slightly from their average equilibrium position creating a macroscopic polarisation that reduces the electric field within the material. Since light is an electromagnetic wave, they create a time-dependent polarisation ($\vec{P}(\omega)$) in the material:

$$\vec{P}(\omega) = \varepsilon_0 \chi_e(\omega) \vec{E}(\omega) \tag{2.11}$$

where $\omega$ is the angular frequency, $\varepsilon_0$ is the electric permittivity of the free space and $\chi_e(\omega)$ is the frequency-dependent susceptibility of the material. The frequency dependency of $\chi_e$ gives rise to the material's chromatic dispersion (i.e. the wavelength dependency of the group and phase velocity for light propagating in the medium). The relative permittivity is related to the susceptibility through:

$$\varepsilon_r = \chi_e + 1 \tag{2.12}$$

For a non-magnetic material, the refractive index $\eta$ is given by Eq. (2.13), which shows how the speed of light is reduced in a medium:

$$\eta = \frac{c_0}{c} = \sqrt{\frac{\varepsilon}{\varepsilon_0}} = \sqrt{1 + \chi_e} \tag{2.13}$$

where $c_0$ is the speed of light in free space and $c$ is the reduced speed of light, also called the group velocity. A refractive index is a dimensionless number describing how light propagates through a material:

$$\eta = \frac{c}{v} \tag{2.14}$$

Snell's law describes how light behaves when passing from one material to another:

$$\frac{\sin\theta_1}{\sin\theta_2} = \frac{v_1}{v_2} = \frac{\lambda_1}{\lambda_2} = \frac{\eta_2}{\eta_1} \tag{2.15}$$

Refraction explains why white light disperses when passing through a prism. Reflection, on the other hand, is either specular (conserving the image) or diffuse (conserving the energy), where the angle of reflection is equal to the angle of incidence.

## Semiconductor Physics pn-Junction

Semiconductors play an essential role when it comes to generating and detecting light. Conductivity is related to the number of charge carriers (electrons) that are free to move inside the material. Metals have many free electrons and, therefore, have high conductivities, while isolating materials have almost no free electrons and near to zero conductivity. Semiconductors have some free electrons and are partially conductive.

Semiconductors function with a highest occupied energy band called the valence band (VB), and a lowest unoccupied energy band called the conduction band (CB). Between these, there is a region of forbidden states called the bandgap, which has a magnitude ($E_g$) defined as the energy difference between the top of the VB and bottom of the CB. Electrons can be excited to the CB (e.g. by absorption of a photon or heat with $E > E_g$). Both the excited electron in the CB and the resulting absence of an electron in the VB (called a hole) are free to move and will contribute to the conductivity. In a pure/intrinsic semiconductor, the number of holes in the VB is equal to the number of electrons in the CB, but this balance can be shifted through the introduction of impurities (i.e. doping the semiconductor). Two differently doped regions adjacent to each other create a junction with different electrostatic potentials.

Impurities that give an abundance of holes are called p-doping and impurities that give an abundance of electrons are called n-doping. A combination of these is called a pn-junction. The potential will separate free charge carriers in the junction; electrons are accelerated towards the n-doped region and the holes towards the p-doped region. An incident photon can excite an electron–hole pair, which is then separated and prevented from recombining. Examples of devices utilising this mechanism are photovoltaic cells and photodiodes. To increase the efficiency of light absorption, an intrinsic region is sometimes placed between the two doped regions to increase the photosensitive area.

## Light Sources and Detection

### Thermal Sources

Incandescent light sources like the sun or old light bulbs emit light as black body radiation (source), which gives a continuous wavelength spectrum with a characteristic peak based on the temperature. At around 798 K (Draper point) all materials give a faint red glow, and at higher temperatures, the colour changes to yellow and becomes whiter as shorter wavelengths become stronger. Tungsten filaments operate below 3695 K (their melting point), and even then, most of the radiated energy lies in the infrared, which explains the low efficiency of incandescent light bulbs.

### Non-thermal Sources

Another way to make light is through luminescence, which is also called cold body radiation. Luminescence photons are emitted as electrons relaxed to a lower energy level in the material, with photon energy equal to the electron energy change. In a semiconductor, the energy levels are called CBs and VBs. Spontaneous emission is a random process and, therefore, the light will be incoherent. The different types are classified according to how the electron was excited to the high energy level initially (e.g. photoluminescence, electroluminescence, chemiluminescence). The light-emitting diode (LED) is an example of electroluminescence, where the electrons and holes are separated by a pn-junction.

Stimulated emission (Einstein) is another process where an incoming photon stimulates the relaxation of an excited photon, producing a second photon with identical frequency and phase. The emitted light is coherent and monochromatic, which are the main requirements of a laser. The second requirement is a reflective cavity that ensures multiple passes of the light beam through the active medium. A partially reflective mirror at one end of the cavity lets some light out of the laser (i.e. the laser output). This balances the exponential growth from the doubling of photons together with losses in the cavity.

## Photodetectors

The detector and its properties decide what can be read out from the sensor. Important parameters are the spectral resolution, response time, responsivity, quantum efficiency, dark current, signal-to-noise ratio (SNR), linearity and stability. The main detectors function is to convert light into an electrical signal and often also from analogue to digital. The most common example is the photodiode, which consists of a semiconductor pn-junction. When light is incident on the diode, a current will flow. For OF applications, pin and avalanche photodiodes are mostly used due to their increased photosensitivity. The semiconductor material used in the diode decides the wavelength range that is absorbed. Typically, Si is used for visible wavelengths and InGaAs for near-infrared applications.

## Spectral Resolution

Ideally, one photon absorbed results in one electron in the output current, but the probability of a photon being absorbed is a function of the photon wavelength. The absorption spectrum describes the spectral responsivity of the photodiode (i.e. the efficacy of current production from incoming photons of different wavelengths). Therefore, the current response does not describe the wavelength of its creator. The spectral resolution can be achieved with a prism or a grating, which spatially resolves the different wavelengths and hence enables one to know which wavelength is incident on the detector. With an array of photodiodes, the different wavelengths can be read out simultaneously; this is called a spectrometer. An optical spectrum analyser (OSA) works the other way around, with one detector and the wavelength being scanned over it. The scanning action is provided either by rotating a grating in a monochromator or through various interferometers (e.g. Fabry–Perot and Michelson). The spectral resolution of an OSA is typically higher than for spectrometers, but the need to scan through the wavelengths makes acquisition slower and the machinery bulky.

# Optical Fibre Waveguides

Optical waveguides are often based on dielectric materials such as silica since the optical losses are minimal. The basis for optical waveguiding can be explained by using Snell's law (Eq. 2.15). For incident light with $\theta_1$ on a dielectric material, the so-called total internal reflection (TIRF) can be obtained if $\theta_2 = 90^{\circ}$. The critical angle $\theta_1 = \theta_c$ for TIRF, and this is calculated:

$$\sin\theta_c = \frac{\eta_2}{\eta_1} \tag{2.16}$$

where $\eta_2$ is the RI of the dielectric material and $\eta_1$ is the RI of the incident light medium. With $\eta_2 > \eta_1$, the optical field will also leak into the medium $\eta_2$. This EM wave is known as the evanescent field. With $\theta_1 > \theta_c$ and $\sin\theta_2 > 1$ we have the E-field in $x$-direction as $E_x = E_0 \exp{(ik_2 \sin \theta_2 x)}$ and the E-field in the $z$-direction into the medium $\eta_2$ as $E_z = E_0 \exp{(ik_2 \cos \theta_2 z)}$. Since $\cos\theta_2 = \sqrt{1 - \sin\theta_2{}^2} = \sqrt{1 - n_2^2/n_1^2}$ becomes a complex number $(i\sqrt{n_2^2/n_1^2 - 1})$, for $\eta_2 > \eta_1$, a decaying E-field in the $z$-direction into medium $\eta_2$ will be observed as $E_z = E_0 \exp{(-\kappa z)}$, where $\kappa$ is the attenuation constant of the evanescent field. The light can be guided in OFs for TIRF.

The RI differences in the OF where $\eta_{\mathrm{cladding}}^2 > \eta_{\mathrm{core}}^2$ creates a TIRF regime. Therefore, in OFs, light is guided efficiently in the IR range due to the small losses in silica. The number of modes guided by the OF is dependent on wavelength, the numerical aperture of the OF ($NA = \sqrt{\eta_{\mathrm{cladding}}^2 > \eta_{\mathrm{core}}^2}$) and the radius of the core. These parameters are composed in a relation called the V-parameter. Small OFs with a core about 8–10 μm (cladding 125 μm) are called single-mode (SM) OFs as one mode is guided in the IR range (1300–1500 nm). Therefore, the E-field is highly confined in the core with evanescent field leaking into the cladding. SM OFs are frequently used in telecommunication applications since they are able to transmit a large amount of information over long distances. By stripping the cladding away from the SM OF it is also possible to use it as a taper where the evanescent field is sensing the RI of the surrounding medium.

Multimode (MM) OFs have larger core sizes than SM OF (around 50–200 μm; cladding 125–300 μm) and guide the light usually as several thousand modes. Therefore, the E-field is less confined in the core than SM OFs, but the NA is larger. Usages of MM OFs are often for short-range telecommunications but also sensor technologies. Since the core is larger for MM OFs, it is easier to handle when stripping away the cladding for evanescent field RI sensing.

SM and MM OFs are presented for sensor applications in this book. OFs might be an advantage in use for sensor technologies as compared to electrical sensors considering that they are immune to electromagnetic interference, are corrosion resistant and possible to miniaturise. The OF sensing can be performed intrinsically or extrinsically referring to sensing external stimuli on the inside or the outside of the OF, respectively. The sensing parameters can, also, be multiplexed spectrally, temporally, or by using polarimetry. This allows multi-point sensing with OFs by using sensor components such as Fibre Bragg Gratings or using pulsed light as distributed OF temperature sensors.

## Intrinsic Optical Fibre Sensors

Intrinsic OF sensors detect light–matter interactions in the interior medium of the optical fibre upon external stimuli from the surrounding medium. Specifically, it is the strain and the compression of the OF that is measured for changes in pressure and temperature of the surrounding medium. For constant temperature and pressure measurements, solid mechanics can also be monitored.

## Discrete Point Temperature Sensors

Fibre Bragg gratings (FBG) are used to measure temperature and strain and are not sensitive to the surrounding medium. They are inscribed into the fibre by illuminating it with fringes of UV light. Because of phase matching between the inscribed RI modulations in the fibre, high reflectivity can be produced as a function of the wavelength:

$$\lambda_B = 2n_e\Lambda \tag{2.17}$$

where $\lambda_B$ is the Bragg wavelength, which is reflected, $n_e$ is the effective refractive index and $\Lambda$ is the grating period. Temperature and strain will affect the grating period and hence the Bragg wavelength $\lambda_B$. The temperature sensitivity of an FBG can be expressed as:

$$\Delta\lambda_B = \lambda_B(\alpha + \xi)\Delta T \tag{2.18}$$

where $\alpha$ and $\xi = dn/dT$ are the thermal expansion coefficient and the thermo optic coefficient, respectively. Temperature sensitivities of 10–16 pm/K have been reported in the literature [1]. This sensor is not sensitive to the surrounding medium because no cladding modes are excited. Multiple gratings with different periods can be placed on the same fibre to read the temperature at each grating location.

To create a micro-optical sensor to be used in electrochemical cells with spatial constraints becomes difficult with regards to light confinement. Two options that appear are to use the fibre as a sensor and to create a waveguide inside the cell. The latter introduces the challenge of coupling light in and out of the sensor.

## Distributed Temperature Sensors

Distributed optical fibre sensors (DFS) are based on the measurement of Brillouin, Rayleigh or Raman scattering of light in the core or core-cladding interface upon external stimuli such as a change in temperature, strain or pressure. The multiplexed sensing is distributed along the length of the fibre with a sensitivity that is proportional to the strength of the scattering. The resolution for the sensing length along the fibre will be inversely proportional to the strength of the scattering.

The spontaneous scattering of light in the optical fibre is either elastic or inelastic. Rayleigh scattering is an elastic scattering process whereas the Brillouin and Raman scattering are inelastic processes. Rayleigh scattering based temperature OF sensors refer to probes that change in the collective oscillation of random fluctuating molecular clouds on a small fraction of the EM frequency and on a small spatial scale.

Different interrogation techniques are utilised to monitor the relevant parameters in order to observe how the scattering processes changes upon external stimuli (e.g. changes in temperature, pressure or strain). The spatial resolution of the measured parameters along the OF is related to the pulse width of the light or the light frequency scanning range guided in the fibre. For optical time-domain reflectometry (OTDR), the spatial resolution on the measured temperature or stress is:

$$\Delta z = \frac{\tau c}{2n_{eff}} \tag{2.19}$$

where $\tau$ is the pulse width and $n_{\text{eff}}$ is the effective RI related to the group RI. The spatial resolution for the optical frequency domain reflectometry (OFDR) is expressed as:

$$\Delta z = \frac{c}{2n_{\text{eff}}\Delta f} \tag{2.20}$$

where $\Delta f$ is the scanning range of the light frequency. A tunable laser is used for the OFDR to scan the $\Delta f$ as well as using the inverse Fourier transformation, recovering the scanned frequency. The local Rayleigh or Brillouin scattering interrogated with short pulses of light has a return intensity dependent on the change in the dielectric parameter $\Delta\varepsilon$, which is a function of the local density $\rho$ and temperature $T$:

$$\Delta\varepsilon = \left(\frac{\partial\varepsilon}{\partial\rho}\right)_T \Delta\rho + \left(\frac{\partial\varepsilon}{\partial T}\right)_\rho \Delta T \tag{2.21}$$

$$\Delta\rho = \left(\frac{\partial\rho}{\partial p}\right)_s \Delta p + \left(\frac{\partial\varepsilon}{\partial s}\right)_p \Delta s \tag{2.22}$$

The variations in the temperature will introduce negligible effects on the dielectric constant; however, the variation in local pressure will change the dielectric constant due to the density fluctuations $\Delta p$ that represents the acoustic waves measured by the Brillouin scattering and the change in entropy $\Delta s$ measured by the Rayleigh scattering.

For Brillouin scattering, it is possible to probe the difference in Stokes and anti-Stokes resonance as a function of the acoustic wave on the OF. The relation for Stokes and anti-Stokes resonance frequency in an OF can be expressed as:

$$\Omega_{\text{Stokes}-\text{anti Stokes}} = 4n_{(\omega)}\omega n_{g(\omega)}\left(\frac{V_a}{c}\right)^2 \tag{2.23}$$

where $\Omega_{\text{Stokes}\,-\,\text{anti Stokes}}$ is the difference between the Stokes and the anti-Stokes Brillouin resonance frequency, $4n_{(\omega)}$ is the RI, $\omega n_{g(\omega)}$ is the group RI, $\omega$ is the light frequency and $V_a$ is the sound velocity. The sound velocity or the acoustic waves of the OF is then probed by measuring the shift in Stokes and anti-Stokes frequency as a function of the change in strain or temperature. For distributed sensing, the Brillouin scattering can be analysed in the frequency, time or the correlation domain.

For the collective oscillation of molecular clouds resulting in a backward propagating Rayleigh scattering, the fluctuating $\Delta\varepsilon(t, z)$ as a function of $\Delta s$, can be probed by correlating it to itself for a known fibre length L. The transverse backward travelling wave of the Ralyeigh scattering in 1D plane wave approximation can be expressed as:

$$E = \psi(z, k) \exp(-ikz) \tag{2.24}$$

where $\psi(z, k)$ is the Rayleigh scattering amplitude that is dependent on the local variations of $\Delta\varepsilon(z)$ within the range $z$. The Fourier transform of the random spatial variation of the permittivity is further related to the Rayleigh backward scattering signal from $\psi(z, k) - \psi(0, k)$. The change in the $\Delta()z$ is then probed by detecting the backward Rayleigh scattering for different $z$ ranges that depend on the change in temperature or strain. The Rayleigh scattering is often analysed either in the OFDR or in the phase OTDR.

In Raman scattering, the change in light–matter interactions with the vibrational modes within the fibre is probed for the change in temperature. Similar to Brillouin scattering, the Raman scattering is also monitored for Stokes and anti-Stokes resonance frequencies. The energy of the light is increased or decreased upon the interactions with the acoustic phonons for Brillouin scattering, whereas, the energy of light in Raman scattering is decreased or increased upon the interactions with the vibrational or rotational modes of a molecule. The exchange of energy between light and molecule will change as a function temperature. Intensities of the spectral line of Stokes and anti-Stokes can be expressed as:

$$\frac{I_{\text{Anti Stokes}}}{I_{\text{Stokes}}} = \left(\frac{\lambda_{\text{Stokes}}}{\lambda_{\text{Anit Stokes}}}\right)^4 \exp\left(\frac{\hbar\omega_M}{k_B T}\right) \tag{2.25}$$

where $I_{\text{Stokes}}$ is the strength of the Stoke line $\lambda_{\text{Stokes}}$, $I_{\text{Anti Stokes}}$ is the strength of the anti-Stoke line $\lambda_{\text{Anit Stokes}}$, $\hbar$ is the reduced Planck constant, $k_B$ is the Boltzmann constant and $T$ is temperature. The temperature can then be monitored by measuring the strength of $I_{\text{Stokes}}$ and $I_{\text{Anti Stokes}}$. The distributed sensing of Raman scattering is usually performed in the OTDR.

## Extrinsic Optical Fibre Sensors

Extrinsic OF sensors detect light–matter interactions in the interface of the OF and the medium. Typically, it is the RI of the surrounding medium that is monitored, but also IR absorption or Raman scattering of specific chemicals can be measured. The RI sensing is performed directly, but entities in the form of molecules in gas or liquid is often measured indirectly. For detection of analytes in gas or aqueous solution, a recognition element is often combined with the OF RI sensor system. The recognition element often is represented by a thin coating of a sensitive material on the OF. Chemical concentrations are generally measured by this method due to their specific binding to a specific thin-coated material.

## Single Point RI or Chemical Optical Fibre Sensors

The extrinsic discrete point OF sensors are similar to those described above but differ in design as the light is interacting with the surrounding medium. The multiplexing capabilities are constrained to spectroscopic measurements of reflected or transmitted light from the periodic inscription of gratings in the fibre. The gratings in the fibre excite cladding modes that are phase sensitive to the surrounding RI. As of yet, there are few discrete multipoint extrinsic OF sensors available in the literature. Several coreless fibres have been spliced together with single-mode fibres to create RI sensing on two or three points [2]. The challenges of combining several RI sensors on the same fibre could be due to challenges concerning amplitude mixing of the several different interference signals. Therefore, this section will mainly focus on the RI sensors available in one parameter configurations as well as the IR absorption and Raman scattering of specific chemicals by the use of OF.

### Long-Period Fibre Grating RI Sensors

Long-period fibre gratings (LPFG) have a grating period longer than the Bragg wavelength, ranging from 100 to 500 μm. The disruption of the core index with a period longer than the Bragg wavelength couples light from the core of the fibre into the modes of the cladding in bands centred at a wavelength $\lambda_n$, $n = 1, 2, 3$ [3]. The scattering losses in the cladding combined with the modes in the cladding interfering with the mode in the core results in dips at $\lambda_n$, $n = 1, 2, 3...$ in the output light. The attenuation bands $\lambda_n$, $n = 1, 2, 3...$ reaction to the grating period and the indexes can be described as:

$$[n_{\mathrm{core}}(\lambda_n) - n_{\mathrm{clad}}(\lambda_n)]\Lambda - \lambda_n = 0 \tag{2.26}$$

where $n_{\mathrm{core}}(\lambda_n)$, $(n_2 < n_{\mathrm{core}} < n_1)$ is the effective index of the core mode, $n_{\mathrm{clad}}(\lambda_n)$, $(n_3 < n_{\mathrm{core}} < n_3)$ is the effective of the $n_{th}$ cladding mode and $\Lambda$ is the grating period [4]. The core mode or $LP_{01}$ of the single mode fibre is insensitive to $n_3$, as the effective index determined from $n_1$ and $n_2$; however, the cladding modes are determined from $n_2$ and $n_3$, where a change in $n_3$ would lead to a change in the coupling wavelength $\lambda_n$.

While the fabrication methods for LPFGs is less complicated than FBGs due to the longer grating period, LPFG based RI sensors will be susceptible to temperature changes. Therefore, a temperature sensor is often combined with the RI monitoring to compensate for changes in temperature due to the LPFG's lack of stability.

### Tilted Fibre Bragg Gratings

An alternative to the TC sensor is the Tilted Fibre Bragg Grating (TFBG) sensor, which eliminates the temperature and refractive index cross-sensitivity inherent to the TC sensor concept. The TFBG sensor can be seen as a combination of a conventional FBG, where periodic density modulations in the core create constructively interfering reflections at a specific wavelength, and a long period grating (LPG), which periodically excites forward propagating cladding modes to interact constructively. In a TFBG, the density modulations are tilted with regard to the fibre axis [5, 6], which causes backwards propagating cladding modes to be excited in addition to the core propagating Bragg resonance [6]. The Bragg wavelength $\lambda_B$ is the same as in Eq. (2.17), but the grating period $\Lambda$ is now the projection of the actual period $\Lambda_g$:

$$\Lambda = \frac{\Lambda_g}{\cos\theta} \tag{2.27}$$

While the Bragg resonance is only temperature sensitive, the cladding modes are also sensitive to RI. The initial approach was to calculate the normalised envelope of the cladding mode resonance spectrum to measure the RI independently [7]. This parameter is relatively insensitive to temperature and is almost linear for RI values between 1.32 and 1.42 [5]. A second option is to compare individual cladding mode resonances with the Bragg resonance, thereby eliminating the temperature effect [6, 7]. Cladding mode resonances can be expressed as:

$$\lambda_C(i) = (n_{\text{eff}}(i) + n_C(i))\Lambda \tag{2.28}$$

where $n_{\text{eff}}$ is the effective RI of the core mode at a wavelength equal to the cladding mode $i$, and $n_C(i)$ is the cladding mode effective RI at this wavelength [8]. The sensitivity of the difference $\lambda_B - \lambda_C(i)$ towards a change in the external medium, RI can be expressed as:

$$\frac{\Delta(\lambda_B - \lambda_C(i))}{\Delta n_{\text{ext}}} = \Lambda \frac{\partial n_C(i)}{\partial n_{\text{ext}}} \tag{2.29}$$

The assumption that the two resonances will have identical temperature shifts has an estimated maximum error of a few pm/K and was measured to 0.5 pm/K [6]. The tilted planes remove the rotational symmetry of the fibre, which introduces a polarisation dependency to the signal. This has been explored further [8]. The benefits of a reflection-based sensor include:

- Reduced size
- Only one entrance necessary in a measurement container
- Reduced cross-sensitivity to strain, compared to a fibre secured in both ends

The following desirable features appear for TFBGs in the reflection configuration [9]:

- **Shifted sensitivity** from RI ranges typically found in glasses (around 1.45), to lower RIs typically found in water-based solutions.
- **Power measurement** instead of wavelength is possible for a lower cost.
- **Self-referencing** to the RI insensitive core mode enables the separation of temperature effects and RI changes that are not caused directly by temperature changes. This can also be used to cancel out power fluctuations in the source.
- **Small size** down to 10–20 mm creates a microprobe that can be inserted in the media of interest.

A few strategies have been investigated to realise a TFBG sensor in reflection mode. By incorporating the TFBG in a section of a TC fibre [10], an RI-independent temperature sensor (TFBG) and a temperature-independent RI sensor (TC) are effectively created. The origin of the RI sensitivity will, in this case, result in increased conversion to propagating modes (i.e. reduced reflected intensity) when the surrounding RI increases. A different solution is to coat the end face with gold, effectively creating a dual-pass transmission TFBG [7]. Offset-splicing has also been used to enhance the back-coupling of cladding modes and thereby creating a power-referenced refractometer [9]. A fourth option is to place a D-shaped core exposed fibre in proximity to the TFBG to absorb the light coupled out; however, this is sensitive to the TFBG rotation angle, as the cladding modes are excited only on one side.

Although the TFBGs are in an early stage of development concerning the fabricating and the characterisation of properties, efforts have been made to apply the TFGB sensor in different applications. A TFBG coated with polyvinyl alcohol have been fabricated in previous work, which showed to detect relative humidity in the range of 20–98%. The sensitivity of this humidity sensor was considerable for high levels of humidity [11].

The specific detection of low molecular weight molecules has also been demonstrated by the use of TFBG coated with an imprinted molecular layer [12]. The limit of detection of maltol is showed to be 1 ng/mL in pure water with a sensitivity of 6.3 x $10^8$ pm/M.

### Infrared Absorption-Based Optical Fibre Sensors

IR absorption spectroscopy with OFs is based on measuring the vibrations of atoms in a molecule to identify and determine concentrations of chemicals. The IR absorption of light occurs at a specific frequency, which can be detected as a dip in the transmission spectrum. The absorption frequency of the IR light corresponds to the energy absorbed by the molecule and is due to the degrees of freedom categorised as translational, rotational or vibrational. The wavenumber as a function of vibrational frequency can be expressed as:

$$\overline{v} = \sqrt{\frac{k}{m}}\left(\frac{1}{2\pi c}\right) \tag{2.30}$$

$$\overline{m} = \frac{m_1 m_2 m_3 \ldots}{m_1 + m_2 + m_3 \ldots} \tag{2.31}$$

where $k$ is the force constant representing the stiffness of the bonds, and $\mu$ is the reduced mass from atoms with mass $m_1,\ m_2,\ m_3 \ldots$. Intuitively, it is the asymmetric vibrations that results in strong absorption caused by the large change in the dipole moment. A specific asymmetric vibration with a specific stiffness $k$ would then correspond to a specific absorption band in the observed spectra.

As molecules exhibit multiple asymmetric vibrational states, the vibrations may take couple over the entire molecule, overlap, or creating overtones, making the spectral analysis complicated and challenging to analyse. Therefore, the frequency regions used for IR absorption are often in the mid-IR for molecular spectroscopy. Fourier-transform IR (FTIR) spectrometers are often used for molecular spectroscopy by the use of Michelson interferometers. Broadband light sources are limited in the mid-IR and narrow linewidth light sources have been used with FTIR spectrometers. Quantum cascade lasers (QCLs) have also emerged as a new platform for transmission spectroscopy that omits the use of the FTIR spectrometers. This simplifies IR absorption spectroscopy significantly as QCLs are tuned to cover a broader bandwidth with high power that allows the IR absorption spectra to be scanned by the light source. Beer–Lamberts law is used to quantify the concentration of the recognised analyte. The absorbance of the sample is related to the concentration as:

$$A = \log \frac{I_0}{I} = \log T \tag{2.32}$$

$$A = \epsilon c l \tag{2.33}$$

where $I_0$ and $I$ are the incident and transmitted light, respectively, $T$ is the transmittance, $\epsilon$ is the absorptivity constant, and $l$ is the pathlength of the light in the sample.

### Raman Spectroscopy-Based Optical Fibre Sensors

Raman spectroscopy-based OF sensors employ Raman scattering to observe vibrational modes associated with specific molecules. This vibrational signature of specific molecules is used to determine their quantity. The signal from Raman scattering is much lower than the signal obtained in detecting IR absorbing molecules. With a low Raman signal, interference may dominate from sources such as fluorescence or Rayleigh scattering. Recent developments in enhancing Raman scattering have led to rapid growth in applying Raman scattering for chemical sensing. The main advantage of using Raman spectroscopy over IR absorption spectroscopy is the excitation wavelength used in the visible or NIR since components in this domain is cheaper, and there is a possibility for generating a broadband light source.

Raman scattering is an inelastic scattering process and can be excited about molecules that change polarisability with no change in dipole when interacting with incident photons. For centrosymmetric molecules, the molecule may be either

IR or Raman active depending on the statement above. A change in dipole and no change in polarisation corresponds to IR active molecule, whereas no change in dipole and change in polarisation corresponds to Raman active molecules.

For Raman active molecules, incident light in the VIS-NIR domain may polarise (distort) the electron cloud around the nuclei, resulting in a short-lived nuclei complex or virtual states between the photon and molecule. This virtual state may also couple with the vibrational energy states in the molecule that induces energy transfers. The energy transfer between the virtual state and the vibrational modes may result in photon scattering with energy smaller or more significant than the excited virtual state, known as Stokes and anti-Stokes Raman scattering, respectively.

The Stoke or anti-Stoke Raman scattering exhibits a wavelength that is lower or higher than the incident photon. This leads to very sharp features in the spectrum due to the shifting of the wavelength in the Raman spectrum and enables flexibility in the choice of the excitation wavelength, avoiding fluorescent interference in the signal.

The coupling and energy transfers between virtual state and vibrational modes happen very rarely, and only one in every $10^6$–$10^8$ photons may result in Raman scattering. It is possible to use surface gratings or nanoparticles of a noble metal or silicon to enhance this process. At a specific resonance light wavelength, the electric field will be large between surface gratings or on the surface of the nanoparticles. The enhancement of the Raman scattering is proportional to the fourth power of the local electric field at Raman site to the incoming electric field ratio. The centre of symmetry can be broken in this process, and some new Raman peaks may appear while other Raman peaks become very weak. The molecule may also become IR absorption active due to the change in centre of symmetry. It is also necessary to employ statistical methods or supervised machine learning for enhanced and non-enhanced Raman scattering peaks to classify chemical components with their respective quantities as for IR absorption spectroscopy.

### Surface Plasmon Resonance-Based Optical Fibre Sensors

Surface plasmonic resonance (SPR) OF sensors exploits the interaction of incident electromagnetic waves with the oscillation electrons on the surface of a metal, where a positive permittivity material shares a boundary with a negative permittivity material [13]. An example of this is the metal gold (which is a negative permittivity material), and palladium (which is a positive permittivity material). When the momentum of incident electromagnetic waves matches the momentum of the surface electron waves in the gold, the electromagnetic waves are coupled to the surface plasmon of the gold at the boundary between the two metals and is diminished. The frequency of the surface electrons can be altered by changes in the material used, resulting in a change in the wavelength of incident electromagnetic waves absorbed. These types of sensors are advantageous, due to the limited complexity, low cost and high specificity. To calculate the absorption at the SPR peak, the following modified Beer–Lambert equation is used:

$$\mathrm{SPR} = -10 \times \log \left\{ \frac{(\text{sensor} - \text{dark})}{(\text{reference} - \text{dark})} \right\} \tag{2.34}$$

where the sensor signal is the transmission signal achieved with the sensor between the light source and the spectrometer, the reference is the transmission signal without the sensor between the light source and spectrometer, and the dark signal is the signal with the light source turned off for both the reference and sensor setup, respectively.

A halogen light source can be used to provide a broad spectrum of visible light, and the signal is straight forward to detect using a spectrometer. When the condition of the SPR is fulfilled, the transmission spectrum of the optical fibre changes due to the SPR-induced spectral absorption peak. This occurs when the longitudinal wave vector of the light reflected at the fibre–gold interface matches the wave vector of the surface plasmonic wave. A layer of $SiO_2$ can use between the layers of gold and palladium (sensing material) to shift the SPR to allow the coupling to the surface plasmonic within the numerical aperture of the OF. The thicker the $SiO_2$ layer is, the larger the angular position and longer the peak wavelength. The thickness of both the gold and palladium layers defines the propagation constant of the surface plasmon.

**Acknowledgements** The authors are grateful to the ENERSENSE programme and NTNU Team Hydrogen at the Norwegian University of Science and Technology (NTNU) for supporting and helping on this book project.

## References

1. Lowder TL, Smith KH, Ipson BL, Hawkins AR, Selfridge RH, Schultz SM (2005) High-temperature sensing using surface relief fiber Bragg gratings. IEEE Photon Technol Lett 17 (9):1926–1928
2. Liu C, Wang F, Lv J, Sun T, Liu Q, Fu C, et al. (2016) A highly temperature-sensitive photonic crystal fiber based on surface plasmon resonance. Opt Commun 359:378–82. Available from: https://doi.org/10.1016/j.optcom.2015.09.108
3. Patrick HJ, Kersey AD, Bucholtz F (1998) Analysis of the response of long period fiber gratings to external index of refraction. J Light Technol 16(9):1606
4. Vengsarkar AM, Lemaire PJ, Judkins JB, Bhatia V, Erdogan T, Sipe JE (1996) Long-period fiber gratings as band-rejection filters. J Light Technol 14(1):58–65
5. Alberto NJ, Marques CA, Pinto JL, Nogueira RN (2010) Three-parameter optical fiber sensor based on a tilted fiber Bragg grating. Appl Opt 49(31):6085–6091
6. Chan C-F, Chen C, Jafari A, Laronche A, Thomson DJ, Albert J (2007) Optical fiber refractometer using narrowband cladding-mode resonance shifts. Appl Opt 46(7):1142–1149
7. Chen C, Yu Y-S, Yang R, Wang C, Guo J-C, Xue Y et al (2012) Reflective optical fiber sensors based on tilted fiber Bragg gratings fabricated with femtosecond laser. J Light Technol 31 (3):455–460
8. Renoirt J-M, Zhang C, Debliquy M, Olivier M-G, Mégret P, Caucheteur C (2013) High-refractive-index transparent coatings enhance the optical fiber cladding modes refractometric sensitivity. Opt Express 21(23):29073–29082
9. Guo T, Liu F, Guan B-O, Albert J (2016) Tilted fiber grating mechanical and biochemical sensors. Opt Laser Technol 78:19–33

10. Gu B, Qi W, Zhou Y, Wu Z, Shum PP, Luan F (2014) Reflective liquid level sensor based on modes conversion in thin-core fiber incorporating tilted fiber Bragg grating. Opt Express 22 (10):11834–11839
11. Miao Y, Liu B, Zhang H, Li Y, Zhou H, Sun H et al (2009) Relative humidity sensor based on tilted fiber Bragg grating with polyvinyl alcohol coating. IEEE Photon Technol Lett 21 (7):441–443
12. Lépinay S, Ianoul A, Albert J (2014) Molecular imprinted polymer-coated optical fiber sensor for the identification of low molecular weight molecules. Talanta 128:401–407
13. Lamb JJ, Bernard O, Sarker S, Lien KM, Hjelme DR (2019) Perspectives of surface plasmon resonance sensors for optimized biogas methanation. Eng Life Sci 19:759–769

# Part II

# Optical Sensor Measurements

# Temperature and Humidity Measurements 3

Markus S. Wahl, Harald I. Muri, Rolf K. Snilsberg, Jacob J. Lamb, and Dag R. Hjelme

In recent years, post-process high-speed, modern signal conditioning methods, low-power and low-cost microelectronic hybrid circuits and advances in miniaturisation technologies have driven improvements in sensor manufacturing [1]. Additionally, the sensor must have a high degree of efficiency relative to its calibration circumstances and sensing mechanism [1]. To save time and improve quality in the implementation of mass production processes, many simulation techniques and design aides can be utilised in the sensor design phase to predict and improve output data [1]. Furthermore, the miniaturisation of sensors offers a wide variety of advantages (e.g. low hysteresis and batch fabrication) [1].

Electrochemical energy systems are subject to the laws of thermodynamics. Therefore, the need for temperature control is paramount. Furthermore, some applications (e.g. water electrolysers and hydrogen fuel cells) require information regarding the water humidity during their function. Both temperature and humidity are essential parameters that need to be measured with regard to understanding the internal environment, characterisation of the performance and optimisation.

The requirements of sensors for use in electrochemical energy systems are that they must be inert, accurate, withstand hazardous conditions and be microscopic. OF sensors provide a platform that can meet all of these requirements for use in

M. S. Wahl · H. I. Muri · R. K. Snilsberg · D. R. Hjelme
Department of Electronic Systems, ENERSENSE, Norwegian University of Science and Technology, Trondheim, Norway
e-mail: markus.s.wahl@ntnu.no; harald.muri@ntnu.no; rolf.k.snilsberg@ntnu.no; dag.hjelme@ntnu.no

J. J. Lamb (✉)
Department of Electronic Systems, ENERSENSE, Norwegian University of Science and Technology, Trondheim, Norway

Department of Energy and Process Engineering, ENERSENSE, Norwegian University of Science and Technology, Trondheim, Norway
e-mail: jacob.j.lamb@ntnu.no



J. J. Lamb, B. G. Pollet (eds.), *Micro-Optics and Energy*,
https://doi.org/10.1007/978-3-030-43676-6_3

electrochemical energy systems. This chapter will cover the current OF temperature and humidity sensors that can be used in electrochemical energy systems.

## Humidity as a Measurable Parameter

Humidity plays a significant role in every part of life on our planet. Be it in biology, or human-engineered processes. In order to maintain the desired humidity, it is essential to monitor, detect and control the humidity. This can be in varied conditions ranging from low to high temperatures or mixtures with other gases [2]. This is also the case in many electrochemical energy systems (e.g. water electrolysers and hydrogen fuel cells).

We have a relational understanding of increased water vapour in daily life as temperature increases or as pressure reduces. Water vapour may also increase due to common chemical reactions such as when our cellular respiration converts $O_2$ and sugar into $CO_2$, water and chemical energy. By monitoring water vapour in addition to monitoring pressure and temperature, we can control thermodynamic processes with much higher accuracy.

### Principle of Humidity Sensing

Humidity is the amount of water in the form of vapour in a body of air or other gases that may follow the gas laws defined by Dalton [3]. The parameters of humidity are defined in a variety of ways with the units of measurement corresponding to the measurement technique used. The most commonly used terms are relative humidity (RH), parts per million (ppm) by weight or by volume and absolute humidity (AB). AB is measured with the unit of grams per cubic metre and is defined as a ratio of the mass of water vapour to the volume of air:

$$\mathrm{AB} = \frac{m_w}{v} \tag{3.1}$$

where AB is the absolute humidity, $m_w$ is the mass of water vapour and $v$ is the volume of air. RH is the ratio of the amount of water vapour relative to the maximum (saturated) water vapour that the air can hold at a given temperature and pressure. Therefore, RH is temperature dependent and is a relative measurement stated as a %:

$$\mathrm{RH}\% = \frac{P_v}{P_s} \times 100 \tag{3.2}$$

where $P_v$ is the actual partial pressure of water vapour in air and $P_s$ is the saturated water vapour pressure of air at the same temperature. Saturation humidity is defined as the ratio of the mass of water vapour at saturation to the volume of air:

$$SH = \frac{m_{ws}}{v} \tag{3.3}$$

where SH is the saturation humidity (g/m$^3$), $m_{ws}$ is the mass of water vapour at saturation (g) and $v$ is the volume of air. Saturation humidity is a function of temperature. It can provide the maximum amount of water vapour content in a volume of gas at a given temperature. RH can also be represented in another way by calculating the ratio of absolute humidity to saturation humidity as a % as follows:

$$RH\% = \frac{AB}{SH} \times 100 \tag{3.4}$$

Parts per million by volume (ppmv) is defined as the volume of water vapour content per volume of dry gas, and parts per million by weight (ppmw) is obtained by multiplying ppmv by the mole weight of water per mole weight of that gas or air. Therefore, ppmv and ppmw are absolute humidity measurements.

## Traditional Optical Humidity Detection

Infrared (IR) optical absorption hygrometers are traditionally used to measure humidity. An IR hygrometers operation is based on a dual-wavelength absorption technique. This uses a primary wavelength and a secondary/reference wavelength. At the primary wavelength, strong optical radiation absorption is observed, and at the secondary/reference wavelength the absorption is negligible. The ratio of transmission between the two wavelengths can then be used as a proxy for humidity measurements. This allows a direct absorption-based measurement of water vapour.

Additionally, the drift and interference caused by contaminants can be minimised. The method's sensitivity depends on the absorption path length and is governed by the Beer–Lambert law [4]. This means that the transmission of the IR radiation through the sample is inversely proportional to the exponential of the concentration and the path length. Despite this instrument having negligible drift with operation over a wide humidity range, they are complex and large measurement solutions.

### Miniaturised Humidity Sensors

Optical waveguide sensors are a class of optical-based device that fall under the category of miniaturised humidity sensors. An example is the surface relief gratings on $SiO_2/TiO_2$ waveguides reported by Tiefenthaler and Lukosz [5]. The roles played by surface and volume adsorption of water molecules were determined quantitatively from the measurements of the effective refractive index changes of specific modes in the waveguide. The observed changes in the refractive index as a result of water adsorption were 0.037 at a wavelength of 514.5 nm, and the number of monolayers of water adsorbed can be determined under different conditions.

**Table 3.1** Fibre optic humidity sensing techniques

| Technique | Method | Range (%RH) | References |
|---|---|---|---|
| Direct spectroscopic | Absorption | 0–95 | [8] |
| | Fluorescence lifetime | 4–100 | [9] |
| | Fluorescence quenching | 0–100 | [10] |
| Evanescent | Absorption | 0–95 | [8] |
| | Attenuation | 25–95 | [11] |
| | Resonance shift | 0–80 | [12] |
| In-fibre grating | Strain-induced | 10–90 | [13] |
| | Resonance shift | 0–95 | [14] |
| Interferometric | Long period grating | 33–97 | [15] |
| | Fabry–Perot intensity | 0–97 | [16] |

With the advent of OF technology, a considerable level of research has been focused on OF-based techniques for humidity sensing. In a similar way to their electronic or mechanical counterparts, OF humidity sensors are secondary devices but show additional features like small size, immunity to electromagnetic interference, multiplexing and remote sensing capabilities, all of which the counterpart electronic sensors lack.

The various OF-based humidity sensing techniques reported so far, with designs covering both extrinsic and intrinsic sensor types, can be further classified according to the techniques commonly employed in OF sensing [6, 7]. Generally, these techniques include direct spectroscopic, evanescent wave, in-fibre grating and interferometric methods, as discussed in detail below (Table 3.1 provides an overview of some selected technologies).

### Direct Spectroscopic Optical Fibre Humidity Sensors

The spectroscopic method has been a "workhorse" technique widely used in chemical analysis. This method examines the optical signal obtained and relates absorption or fluorescence-based measurements to the concentration of the target analyte. Hence, it is not surprising that it is a popular choice of method for many OF-based chemical sensors. Generally, the design of the sensors can comprise optical fibres with a sample cell for direct spectroscopic measurements or be configured as fibre optrodes where a chemical selective layer, comprising chemical reagents in suitable immobilising matrices, is deposited onto the optical fibre [6, 17]. Most spectroscopic-based configurations for humidity sensing are based on the optrode design where moisture-sensitive reagents are attached to the tip of the sensing fibre, usually with the aid of a polymeric material to form the supporting matrix.

Absorption-based optical fibre humidity sensors are another form of direct spectroscopic technique. A variety of potential materials and chemical reagents have been reported in light of their humidity-dependent optical absorption properties. Examples include reagents such as cobalt chloride ($CoCl_2$), cobalt oxide ($Co_3O_4$), Rhodamine B, crystal violet, and more recently, materials such as electrochromic polymers and bacteriorhodopsin doped bio-chromic film [18]. With these chemicals, measurements are made by monitoring the intensity variation as a result of absorption due to the interaction between the chemical reagents involved and moisture. These materials can be applied to the OF by having a porous section of the fibre that is doped with the chemical, use of sol–gel techniques, or using an air-gap design.

Many fluorescence-based sensors can also be classed as direct spectroscopic humidity sensors. A humidity-sensitive optrode membrane fabricated using a dye-doped (Rhodamine 6G) gelatine film has been observed to function as an OF-based humidity sensor [19]. The sensing membrane, which can be readily adapted for OF sensing, exhibits strong fluorescence at 568 nm when excited at 536 nm. A hypochromic shift was observed with a resultant decrease in the emission intensity at 568 nm when the humidity level was increased from 0 to 100% RH.

### Evanescent Optical Fibre Humidity Sensors

The evanescent wave (EW) sensing method allows the OF to be used as an intrinsic sensor where the field generated at the interface interacts with the target analyte surrounding the fibre, giving information as a result of optical absorption, refractive index change or scattering [6, 18]. Various OF configurations can be used for evanescent wave sensing. A common approach is to use a plastic-clad OF with a section of the cladding removed, in order to gain access to the evanescent field. The structure of the de-clad fibre can also be modified by heating and bending it to form a U-bent fibre, causing the evanescent field to extend further away from the interface, hence enhancing the interaction between light and the target analyte. Other methods to gain access to the evanescent field include side polishing of optical fibres cast in a polymer block to expose the fibre core or heating and stretching the optical fibre to form a fibre taper.

Chemical reagents or selected matrices can be coated onto a de-clad OF, forming a standard design configuration for EW-based sensors. Comparing these to the sensors discussed in the previous section, the use of such a configuration gives flexibility in terms of the interaction length, time response and distributed sensing capability [6, 18]. However, the required optical path length to achieve a similar sensitivity as is achieved for example in the direct absorption method either by using a direct sample cell or configuration based on active fibre core is much longer due to the small path length interaction at every reflection point along the fibre [6].

RI change is another approach frequently used in the EW sensing method. An example of humidity sensing based on this method has been demonstrated using a plastic OF [20]. The fibre core of the plastic OF (diameter: 1 mm) was made from PMMA with a refractive index of 1.489 at 680 nm. Hence to render the fibre responsive to humidity, a polymer blend of HEC/polyvinylidene fluoride (PVDF) was deposited on the fibre core, forming a cladding layer (thickness: 0.5–1 μm). The humidity-sensitive cladding layer has a refractive index value of 1.492 when in a dry state, creating a lossy waveguide which reduces the intensity of the light propagating through the fibre. As the cladding layer hydrates, the refractive index value falls below that of the core, reducing the intensity loss, forming the basis for humidity detection.

### In-Fibre Grating Humidity Sensors

The in-fibre grating sensor represents a class of intrinsic OF sensors that have gained widespread popularity in recent decades. They have been used in numerous applications in various industries due to its inherent sensitivity to temperature, strain and refractive index change [18]. The grating structure within the fibre sensor is created by UV-induced periodic refractive index modulation of the fibre core and can be generally classified into two main categories depending on the grating period, namely the fibre Bragg grating (FBG) and long-period grating (LPG).

Grating-based sensors are commonly used in chemical sensing. The LPG can be employed as a general refractive index sensor and used in conjunction with chemical selective materials to create a species-specific chemical sensor. This thus forms a desirable refractive index-based chemical sensing mechanism which has been employed in the detection of a variety of chemical species [18]. The FBG-based sensors, on the other hand, are on the whole used predominantly for the monitoring of physical parameters such a temperature, strain or pressure. To use an FBG-based sensor as a chemical sensor, a common approach is to select a material selectively sensitive to the chemical measurand and capable of inducing mechanical deformation as it interacts. This is done to produce a secondary strain-induced measurement using the FBG sensor as a result of physical or chemical interactions. The selection of the sensing materials for an FBG chemical sensor is, therefore, more stringent than LPG as it should be not only responsive to the selected chemical species but also be able to expand in order to induce strain on the FBG. Applications of FBGs in chemical sensing reported so far are primarily limited to the essential fields of $H_2$ gas detection, salinity measurement and moisture sensing [18].

### Interferometric Optical Fibre Humidity Sensors

Optical interferometry is a powerful and versatile tool that has been applied in optical fibre sensing to yield high performance OF sensors. In addition to the advantages attributed to the use of fibre optics, OF interferometric sensors generally provide geometric versatility in terms of sensor design and a high level of measurement sensitivity [21]. The sensing mechanism relies on the perturbation of the phase properties of the light signal travelling in the optical fibre introduced by an external environment. The detection of the phase change is realised by mixing the signal of interest with a reference signal, consequently converting the phase difference between the two signals into an optical intensity change. Various interferometer configurations such as the Mach–Zehnder, Michelson, Sagnac and Fabry–Perot can be used to perform the detection.

## Current Optical Temperature Sensor Technologies

There are a wide variety of applications of fibre optic sensors in composite materials, including temperature measurements. For this application, the critical requirement is to measure either strain or temperature, or both parameters simultaneously. In this section, a discussion of the different types of OF sensors that can be used with composite materials to measure strain/temperature when embedded inside a system is made. The different types of OF sensors reported for strain/temperature measurements in composite materials are fibre Bragg grating (FBG) sensors, interferometric fibre optic sensors, polarimetric sensors, fibre optic micro bend sensors, distributed sensors (using techniques such as Rayleigh scattering, Raman scattering and Brillouin scattering) and hybrid sensors [22].

Many strategies to measure temperature with a fibre optic sensor have been developed since Corning developed the first low-loss optical fibre in 1970 [23]. As described in Chap. 2, fibre optic sensors can be classified as either intrinsic or extrinsic. Although many extrinsic sensors are temperature sensitive, their inherent cross-sensitivity to other parameters excludes them from general use. An interesting exception is the tilted fibre Bragg grating (TFBG) sensors, with their intrinsic temperature sensitivity and extrinsic RI sensitivity. Discrimination between parameters is necessary with one sensor that provides two parameters with different responses (e.g. fundamental and second-order Bragg resonance).

## Blackbody Radiation-Based Temperature Sensing

The blackbody radiation temperature measurement method is based on the thermal radiation principle [24–28]. Any object whose temperature is higher than absolute zero will emit infrared radiation to the outside. The infrared radiation energy of the object and the distribution of its wavelength are closely related to its surface temperature. Therefore, the surface temperature of the object can be accurately measured by measuring the infrared energy of its radiation. A detection fibre, which is a crystallised fibre form by annealing twice at high temperature, is fused to the end of transmission fibre. The thermal radiation energy of the high-temperature field will be coupled to the end surface of the transmission fibre by a detection fibre and then transmitted to the spectral analyser. According to Planck's radiation law, the change of temperature can be derived from the light intensity of the thermal radiation [24, 28, 29]. Generally, OF sensors based on blackbody radiation are reserved for use in high-temperature applications above 400°C.

## Absorption-Based Temperature Sensing

Tuneable diode laser absorption spectroscopy (TDLAS) is a well-used technique for OF temperature sensing and has been used widely for the detection of temperatures and temperature distributions [30]. The system is based on a laser beam path that can be detected using a spectral sensor. The spectral information can then be used to calculate the temperature between the laser diode and the spectral sensor using the Lambert–Beer law. By integrating a matrix of laser beams and spectral detectors, a multidimensional distribution of temperature can be observed.

## Polarimetric-Based Temperature Sensors

The polarisation properties of light propagating through an optical can be affected by stress, strain, pressure and temperature acting on a measuring fibre and, in a fibre, polarimetric sensor, the polarisation change is detected to retrieve the sensing parameter [22]. Polarimetric sensors detect the presence of some physical field via a change in the birefringence of the sensing fibre. The transduction mechanism is the modulation of the polarisation state or the optical phase difference between the two orthogonal eigenstates. Such sensors are therefore analogous to interferometric sensors where the resultant intensity, arising from the superposition of two mutually coherent light beams corresponding to two different paths, varies with the relative path difference with a period equal to the optical wavelength [31]. The polarimetric sensor demonstrated here utilises the orthogonal birefringent axes of polarisation maintaining fibre to provide two independent paths within the fibre. If both modes are equally excited, providing that the two polarisation modes can mix coherently, an interference pattern of maximum visibility will be observed upon modulation of the fibre birefringence [31].

## Interferometer-Based Temperature Sensors

Interferometric fibre sensors can also be employed for strain/temperature measurements in electrochemical energy systems. There are different types of interferometric fibre sensors which differ in their operating principles and strain/temperature sensing characteristics [32]. The interferometric fibre sensors most commonly employed for sensing applications are extrinsic Fabry–Perot interferometers (EFPI), and micro-hole collapsed modal interferometers.

A Fabry–Perot interferometer (FPI) generally comprises two parallel reflecting surfaces separated by a certain distance [33]. It comprises a superluminescent light diode (SLD) source, a coupler and a spectrometer. Any applied longitudinal strain to the FPI sensor alters the physical length of the cavity, which results in a phase difference between reflected or transmitted beams. By measuring the shift of the wavelength spectrum, the strain applied to the FPI can be measured. It is found that the shorter the optical path difference (OPD), the larger will the free spectral range (FSR) be, resulting in a wider dynamic range for a sensor [34]. Therefore, the dynamic range of the sensor can be tuned by varying the cavity length, which in turn changes the OPD of the FPI sensor [35].

A large group of temperature sensors are based on modal fibre interferometers. Higher-order modes are excited in the sensitive section of the fibre, and the difference in effective refractive index creates interference between them that is sensitive to temperature [36]. The thermal sensitivity stems from thermal expansion and the thermo-optic effect, which both affect the optical path length ($OPL = nd$) of the modes. If the higher-order modes are cladding modes, or in other way reach the fibre surface, the sensor will have a cross-sensitivity to the refractive index of the surrounding medium. Efforts have been made to untangle the two parameters

[37, 38], but the method relies on linear response to both parameters. Lee et al. (2012) provide an in-depth review of interferometric fibre optic sensors [33]. The sensitivity of interferometer-based sensors is observed as a peak wavelength change and range from 15 pm per °C for multimode interferometers [39], 58.5 pm per °C for gradient index-multimode interferometers [40] and 81.7 pm per °C for double taper interferometers [41].

## Fibre Bragg Grating Temperature Sensors

Fibre Bragg grating (FBG)-based temperature sensors are one of the most promising solutions, giving good sensitivity and a robust configuration. An FBG sensor comprises a grating region with a periodic change in refractive index in the core region of an optical fibre. Such periodically modulated refractive index structures enable the light to be coupled from the forward propagating core mode into a backward propagating core mode generating a reflection response. Chapter 2 describes the working principle and fabrication in detail. FBGs utilise the photosensitive property of the optical fibre material to establish a spatial periodic refractive index distribution on the core of the fibre (by the inscribed grating). This results in a change or allows the control of the propagation behaviour of the light. As the grating is inscribed into the fibre itself, the integrity of the fibre is not compromised, although the hydrogen loading does make the glass material more brittle [42]. The glass is inert to most substances and is therefore ideal to use in harsh environments, with high-temperature resolutions [43].

Commonly FBGs are used to measure axial strain as well as temperature. This is because axial strain sensitivity is effectively higher as the change in the FBG pitch is directly proportional to the applied longitudinal strain. When there are changes in the external temperature or strain, there will be changes in the period of the FBG resulting in changes to the refractive index of the core. This causes the Bragg wavelength of the FBG to change. The Bragg wavelength follows these changes, which is the basic principle of a wavelength modulation sensor like an FBG sensor. This means that as long as the Bragg wavelength change can be measured, the strain and temperature can be obtained. Therefore, calibration of FBGs for the desired temperature is most important. The temperature sensitivity of an FBG is observed as a peak wavelength change of 10 pm per °C [44]. This allows for a dynamic range of 0–200°C in conventional FBG sensors [45], and 0–400°C in more advanced systems [46].

### Long-Period Fibre Grating

Long-period fibre gratings (LPFGs) have attracted much attention because they can form many useful devices, including optical sensors. An LPFG can be produced by introduction of a periodic modulation of the refractive index with a pitch of the order of 100 μm along with the core of a single-mode fibre [47]. The transmission spectrum of a typical LPFG consists of several rejection bands that arise from light coupling from the fundamental mode to the cladding modes of the fibre. In general, these rejection bands are sensitive to the temperature, the strain along the LPFG and the refractive index of the surrounding medium. Measurement of the shift of the rejection band in response to a change in these physical parameters is the basis of LPFG sensors. Separation of various physical effects is, therefore, an essential issue in the development of practical LPFG sensors. For refractive index measurement, in particular, the temperature effect must be eliminated. The temperature sensitivity of an LPFG can be reduced with a specially doped host fibre [48] or a carefully chosen polymer coating.

### Tilted Fibre Bragg Grating

Recently, tilted fibre Bragg gratings (TFBGs) have been considered as the critical components in novel fibre sensors [49–55] because they have the sensing characteristics of both uniform fibre Bragg gratings (FBGs) and long-period fibre gratings (LPFGs) [55]. In the TFBGs, light coupling occurs not only between the forward and backward core modes but also between the forward core mode and backward cladding modes and even radiation modes [56]. Therefore, it is an ideal element for constituting a reflective fibre device based on cladding-mode recoupling. Several reflective TFBG devices have been proposed after introducing some unique cladding-mode exciting or recoupling element [57]. This allows the production of multiparameter sensors for simultaneous measurement of strain, temperature and RI [50]. The temperature sensitivity of a TFBG is observed as a peak wavelength change of 14.2 pm per °C, allowing for a broad dynamic range of 0–800°C [58].

## Some Challenges and Solutions for Optical Fibre-Based Sensing

Embedding OF sensors inside electrochemical energy systems is a minimally invasive technique, but for industrial applications still, there exist a few issues that are under investigation. One of the challenges is the provision of a reliable method for connecting to the sensor [22]. Custom designed ingress/egress optical fibre end connectors show promise; however, these may cause the fibre to become brittle near the edges of the structure [22]. Furthermore, after the connector installation, it is impossible to modify the fibre. Free space coupling can be considered a more promising method [22]. A novel example of free-space passive coupling of light into FBG sensors has been shown consisting of a 45° mirror integrated directly into the fibre [59].

The second issue is the structural damage. Since optical fibres can have larger diameters when compared to reinforcement fibres, damages can occur. One way to overcome this is to reduce the diameter of the optical fibres further and apply an optimised protective coating. An example is the Draw Tower Grating (DTG) fibre with an 80 μm diameter. This should be a less invasive optical fibre [60].

The OF sensor embedding procedure for mass production of systems with OF sensors is often a labour-intensive task. Ideally, a reliable automated optical fibre placement system is required for mass production solutions, that can be retrofitted into the existing industrial production processes. Currently, the security of sensitive areas of OF sensor (e.g. grating written area of FBG) and control over the alignment of specialised OF sensors are not well assured. An option to overcome this is an automated fibre placement system combined with X-ray-based micro-controlled tomography. This could be a suitable solution for quality control of the embedded FOS.

Finally, the limitations of the operating range and accuracy of the OF-based sensors are some of the drawbacks. Moreover, the stability and detection range of these sensors needs to be extended to improve their utilisation. The ability for the sensors to withstand harsh environments is also a crucial parameter, and this can come at a cost to the sensitivity of the sensor. Nevertheless, these sensors have found useful applications in various areas where electronic sensors were found to be inappropriate, thereby showing the real potential of OF-based sensors. This forms the primary motivation that has driven the research activities on the development of a range of OF sensors over the last decades, and a lot of these developments can be directly applied to detection in the electrochemical energy systems.

**Acknowledgements** The authors are grateful to the ENERSENSE programme and NTNU Team Hydrogen at the Norwegian University of Science and Technology (NTNU) for supporting and helping on this book project.

## References

1. Farahani H, Wagiran R, Hamidon M (2014) Humidity sensors principle, mechanism, and fabrication technologies: a comprehensive review. Sensors 14(5):7881–7939
2. Carr-Brion K (1986) Moisture sensors in process control. Elsevier, London
3. Wiederhold PR (2012) Water vapor measurement: Methods and instrumentation. CRC, Boca Raton
4. Willard HH, Merritt Jr LL, Dean JA, Settle Jr FA (1988) Instrumental methods of analysis. Wadsworth, Belmont, CA
5. Tiefenthaler K, Lukosz W (1985) Grating couplers as integrated optical humidity and gas sensors. Thin Solid Films 126(3–4):205–211
6. Grattan KTV, Meggitt BT (1995) Optical fiber sensor technology, vol Vol. 1. Springer, Dordrecht
7. Grattan KTV, Sun T (2000) Fiber optic sensor technology: an overview. Sensors Actuat A Phys 82(1–3):40–61
8. Otsuki S, Adachi K, Taguchi T (1998) A novel fiber-optic gas sensing arrangement based on an air gap design and an application to optical detection of humidity. Anal Sci 14(3):633–635
9. Bedoya M, Díez MT, Moreno-Bondi MC, Orellana G (2006) Humidity sensing with a luminescent Ru (II) complex and phase-sensitive detection. Sensors Actuat B Chem 113(2):573–581
10. Posch HE, Wolfbeis OS (1988) Optical sensors, 13: fibre-optic humidity sensor based on fluorescence quenching. Sensors Actuat 15(1):77–83
11. Ogawa K, Tsuchiya S, Kawakami H, Tsutsui T (1988) Humidity-sensing effects of optical fibres with microporous SiO/sub 2/cladding. Electron Lett 24(1):42–43
12. Alvarez-Herrero A, Guerrero H, Levy D (2004) High-sensitivity sensor of low relative humidity based on overlay on side-polished fibers. IEEE Sensors J 4(1):52–56
13. Kronenberg P, Rastogi PK, Giaccari P, Limberger HG (2002) Relative humidity sensor with optical fiber Bragg gratings. Opt Lett 27(16):1385–1387

14. Luo S, Liu Y, Sucheta A, Evans MK, Van Tassell R (2002) Applications of LPG fiber optical sensors for relative humidity and chemical-warfare-agents monitoring. In: Advanced sensor systems and applications. International Society for Optics and Photonics, Bellingham, WA, pp 193–204
15. Venugopalan T, Yeo TL, Sun T, Grattan KTV (2008) LPG-based PVA coated sensor for relative humidity measurement. IEEE Sensors J 8(7):1093–1098
16. Yu H, Yao L, Wang LX, Hu WB, Jiang DS (2001) Fiber optic humidity sensor based on self-assembled polyelectrolyte multilayers. J Wuhan Technol Mater Sci Ed 16(3):65–69
17. Gauglitz G (1996) Opto-chemical and opto-immuno sensors. Sensors Update 1(1):1–48
18. Yeo TL, Sun T, Grattan KTV (2008) Fibre-optic sensor technologies for humidity and moisture measurement. Sensors Actuat A Phys 144(2):280–295
19. Choi MMF, Tse OL (1999) Humidity-sensitive optode membrane based on a fluorescent dye immobilized in gelatin film. Anal Chim Acta 378(1–3):127–134
20. Muto S, Suzuki O, Amano T, Morisawa M (2003) A plastic optical fibre sensor for real-time humidity monitoring. Meas Sci Technol 14(6):746
21. Dandridge A (1991) Fiber optic sensors based on the Mach-Zehnder and Michelson interferometers. Fiber Opt Sensors Introd Eng Sci:271–323
22. Ramakrishnan M, Rajan G, Semenova Y, Farrell G (2016) Overview of fiber optic sensor technologies for strain/temperature sensing applications in composite materials. Sensors 16 (1):99
23. Kapron FP, Keck DB, Maurer RD (1970) Radiation losses in glass optical waveguides. Appl Phys Lett 17(10):423–425
24. Wei W, Xiaotian S, Ying W (2015) Sapphire fiber-optic temperature sensor based on black-body radiation law. Procedia Eng 99:1179–1184
25. Masek V, Mojzes P, Palacky J, Bok J, Anzenbacher P (2010) Binding of platinum complexes to DNA monitored by Raman spectroscopy. In: AIP conference proceedings. AIP, New York, NY, pp 416–417
26. Guo Y, Xia W, Hu Z, Wang M (2017) High-temperature sensor instrumentation with a thin-film-based sapphire fiber. Appl Opt 56(8):2068–2073
27. Pan B, Hao X, LI W, Zhou H (2011) Application of the SSPM in sapphire Fiber black-body cavity transient high temperature sensor. J North Univ China Nat Sci Ed 32(5):619–624
28. Wähmer M, Anhalt K, Hollandt J, Klein R, Taubert RD, Thornagel R et al (2017) Thermodynamic temperature of high-temperature fixed points traceable to blackbody radiation and synchrotron radiation. Int J Thermophys 38(10):144
29. Ogarev SA, Khlevnoi BB, Samoilov ML, Otryaskin DA, Grigor'eva IA, Solodilov MV et al (2016) High-temperature blackbody models for use in photometry, radiometry, and radiation thermometry. Meas Tech 58(11):1255–1260
30. Bolshov MA, Kuritsyn YA, Romanovskii YV (2015) Tunable diode laser spectroscopy as a technique for combustion diagnostics. Spectrochim Acta B 106:45–66
31. Mondanos M, Giles I, Weir K (2005) Fibre optic polarimetric temperature sensor using low coherence source employing intensity and wavelength compensation. In: 17th international conference on optical fibre sensors. International Society for Optics and Photonics, Bellingham WA, pp 647–650
32. Rao Y-J (2006) Recent progress in fiber-optic extrinsic Fabry–Perot interferometric sensors. Opt Fiber Technol 12(3):227–237
33. Lee BH, Kim YH, Park KS, Eom JB, Kim MJ, Rho BS et al (2012) Interferometric fiber optic sensors. Sensors 12(3):2467–2486
34. Udd E, Spillman Jr WB (2011) Fiber optic sensors: an introduction for engineers and scientists. Wiley, Hoboken, NJ
35. Liu T, Wu M, Rao Y, Jackson DA, Fernando GF (1998) A multiplexed optical fibre-based extrinsic Fabry-Perot sensor system for in-situ strain monitoring in composites. Smart Mater Struct 7(4):550
36. Chen C (2009) Sensing characteristics of core and cladding modes in conventional single mode fibre and photonic crystal fibre. Carleton University, Ottawa

37. Li L, Xia L, Xie Z, Hao L, Shuai B, Liu D (2012) In-line fiber Mach–Zehnder interferometer for simultaneous measurement of refractive index and temperature based on thinned fiber. Sensors Actuat A Phys 180:19–24
38. Xiong R, Meng H, Yao Q, Huang B, Liu Y, Xue H et al (2014) Simultaneous measurement of refractive index and temperature based on modal interference. IEEE Sensors J 14(8):2524–2528
39. Li E, Wang X, Zhang C (2006) Fiber-optic temperature sensor based on interference of selective higher-order modes. Appl Phys Lett 89(9):91119
40. Liu Y, Wei L (2007) Low-cost high-sensitivity strain and temperature sensing using graded-index multimode fibers. Appl Opt 46(13):2516–2519
41. Wei T, Lan X, Xiao H (2009) Fiber inline core–cladding-mode Mach–Zehnder interferometer fabricated by two-point $CO_2$ laser irradiations. IEEE Photon Technol Lett 21(10):669–671
42. Martelli C, Mendez A, Triques ALC, Braga AMB, Canning J, Cook K et al (2011) Impact of hydrogen-induced effects on optical fiber Bragg gratings. In: 21st international conference on optical fiber sensors. International Society for Optics and Photonics, Bellingham, p 775385
43. Lowder TL, Smith KH, Ipson BL, Hawkins AR, Selfridge RH, Schultz SM (2005) High-temperature sensing using surface relief fiber Bragg gratings. IEEE Photon Technol Lett 17 (9):1926–1928
44. Guo T, Liu F, Guan B-O, Albert J (2016) Tilted fiber grating mechanical and biochemical sensors. Opt Laser Technol 78:19–33
45. Erdogan T, Mizrahi V, Lemaire PJ, Monroe D (1994) Decay of ultraviolet-induced fiber Bragg gratings. J Appl Phys 76(1):73–80
46. Rathje J, Kristensen M, Pedersen JE (2000) Continuous anneal method for characterizing the thermal stability of ultraviolet Bragg gratings. J Appl Phys 88(2):1050–1055
47. Vengsarkar AM, Lemaire PJ, Judkins JB, Bhatia V, Erdogan T, Sipe JE (1996) Long-period fiber gratings as band-rejection filters. J Lightwave Technol 14(1):58–65
48. Shima K, Himeno K, Sakai T, Okude S, Wada A, Yamauchi R (1997) A novel temperature-insensitive long-period fiber grating using a boron-codoped-germanosilicate-core fiber. In: Proceedings of optical fiber communication conference. IEEE, Dallas, TX, pp 347–348
49. Chan C-F, Chen C, Jafari A, Laronche A, Thomson DJ, Albert J (2007) Optical fiber refractometer using narrowband cladding-mode resonance shifts. Appl Opt 46(7):1142–1149
50. Alberto NJ, Marques CA, Pinto JL, Nogueira RN (2010) Three-parameter optical fiber sensor based on a tilted fiber Bragg grating. Appl Opt 49(31):6085–6091
51. Shevchenko YY, Albert J (2007) Plasmon resonances in gold-coated tilted fiber Bragg gratings. Opt Lett 32(3):211–213
52. Wong ACL, Chung WH, Tam H-Y, Lu C (2011) Single tilted Bragg reflector fiber laser for simultaneous sensing of refractive index and temperature. Opt Express 19(2):409–414
53. Zhou B, Zhang AP, He S, Gu B (2010) Cladding-mode-recoupling-based tilted fiber Bragg grating sensor with a core-diameter-mismatched fiber section. IEEE Photon J 2(2):152–157
54. Villanueva GE, Jakubinek MB, Simard B, Oton CJ, Matres J, Shao L-Y et al (2011) Linear and nonlinear optical properties of carbon nanotube-coated single-mode optical fiber gratings. Opt Lett 36(11):2104–2106
55. Laffont G, Ferdinand P (2001) Tilted short-period fibre-Bragg-grating-induced coupling to cladding modes for accurate refractometry. Meas Sci Technol 12(7):765
56. Erdogan T, Sipe JE (1996) Tilted fiber phase gratings. JOSA A 13(2):296–313
57. Lamb JJ, Bernard O, Sarker S, Lien KM, Hjelme DR (2019) Perspectives of surface plasmon resonance sensors for optimized biogas methanation. Eng Life Sci 19:759–769
58. Chen C, Yu Y-S, Yang R, Wang C, Guo J-C, Xue Y et al (2012) Reflective optical fiber sensors based on tilted fiber Bragg gratings fabricated with femtosecond laser. J Lightwave Technol 31 (3):455–460
59. Qiu L, Goossen KW, Heider D, O'Brien DJ, Wetzel ED (2010) Nonpigtail optical coupling to embedded fiber Bragg grating sensors. Opt Eng 49(5):54402
60. Chojetzki C, Rothhardt MW, Ommer J, Unger S, Schuster K, Mueller H-R (2005) High-reflectivity draw-tower fiber Bragg gratings—arrays and single gratings of type II. Opt Eng 44(6):60503

# Hydrogen Gas Measurements

4

Harald I. Muri, Jacob J. Lamb, Markus Wahl, Rolf K. Snilsberg, and Dag R. Hjelme

To realise the goal of high efficiency, low-pollution, sustainable and abundant energy sector, hydrogen will play a significant role in energy storage [1]. Gas composition measurements for $H_2$ energy storage applications mainly consist of detecting gas concentrations such as $H_2$ to evaluate the electrochemical processes as well as their performance. The measurement of one or more gas components and concentrations may serve as important indicators for the state of chemical or thermodynamic processes in energy systems such as mass transport in fuel cells, water electrolysers or concentration batteries. Gas compositions are usually measured by using IR light absorption sensors or Raman scattering sensors and can be combined with optical fibres to confine the molecule detection at a micro-scale. This chapter presents principles for sensing different gas components, possible types of optical fibre systems solutions for gas detection and challenges and advantages with these.

H. I. Muri · M. Wahl · R. K. Snilsberg · D. R. Hjelme
Department of Electronic Systems, ENERSENSE, Norwegian University of Science and Technology, Trondheim, Norway
e-mail: harald.muri@ntnu.no; markus.s.wahl@ntnu.no; rolf.k.snilsberg@ntnu.no; dag.hjelme@ntnu.no

J. J. Lamb (✉)
Department of Electronic Systems, ENERSENSE, Norwegian University of Science and Technology, Trondheim, Norway

Department of Energy and Process Engineering, ENERSENSE, Norwegian University of Science and Technology, Trondheim, Norway
e-mail: jacob.j.lamb@ntnu.no



J. J. Lamb, B. G. Pollet (eds.), *Micro-Optics and Energy*,
https://doi.org/10.1007/978-3-030-43676-6_4

## Traditional Gas Optical Measurements

Gas measurements with optical systems are based on using light to detect the translational, rotational or vibrational stretching or bending modes that are associated with a specific gas molecule. These vibrational modes may be symmetric or asymmetric, as shown in Chap. 2. Symmetric vibrational modes are usually weaker than asymmetric vibrational modes. For centrosymmetric molecules, a change in the dipole occurs while no change in polarisation is observed when probing with IR absorption methods. For Raman scattering molecules there is no change in dipole, but a change in polarisation is observed.

### Infrared Absorption

For IR absorption, the light wavelengths between 700–2500 nm (IR) and 2500–15,000 nm (MIR) have photon energy that matches the difference in the vibrational stretching or bending energy modes of a gas molecule. When this energy mode matches the photon energy, the molecule absorbs that wavelength of light. The different vibrational modes are the fingerprint for a specific gas molecule, and an IR light of different wavelengths can be used to detect it. Despite this, some molecules have very similar fingerprints, and the IR absorption wavelengths may overlap. This can make the detection of specific gas molecules difficult. The different IR absorption wavelengths can be identified with their respective molecules by using standard chemometric statistical methods. The amount of light absorbed by a specific molecule is related to its concentration in a sample volume by using the Beer–Lambert law as described in Eqs. (2.31) and (2.32).

### Raman Scattering

For Raman scattering, light wavelengths in VIS or IR spectrums may interact with the molecule by initially causing an oscillating polarisation or virtual state. This can couple to vibrational modes that result in an energy transfer between the virtual state and the vibrational modes. As a result of the energy transfer, the scattered light has a wavelength shift increase or decrease that is known as Stokes or anti-Stokes Raman scattering, respectively. The magnitude of the wavelength shift of the scattered light is directly proportional to the concentration of the Raman active molecule. However, the Raman scattering intensity is very weak as it only occurs with every $10^6$–$10^8$ scattered photons.

## Raman- and IR-Based Optical Fibre Hydrogen Sensors

Gas composition measurements with optical fibres require light interaction with the surrounding environment. To achieve this, it is possible to use different types of extrinsic optical fibres, as mentioned in Chaps. 2 and 3. The simplest way to detect gas is to use optical fibre systems in transmission or reflection.

For IR active molecules, it is possible to use single-wavelength light for detecting the target gas. This requires no overlap between absorption bands of target gas and other absorption bands with other gases present in the measurement cavity. In energy storage systems, this is often not the case, and many gases are present with overlapping absorption bands. To separate the absorption bands, it is possible to use a wideband wavelength source so an absorption spectrum can be obtained. Furthermore, the deconvolution method can be applied to the spectra to find the overlapping absorption amplitudes. The light sources able to produce a wideband IR and MIR spectra is often based on quantum cascade lasers. However, to date, these systems are costly to make and are mostly used in dedicated research laboratories. The cheapest and most common method for obtaining broad IR spectra is to use an FTIR spectrometer. This method requires sufficient computation power due to the Fourier transform applied on the interferogram produced by the Michelson interferometer.

It is also possible to excite the light into the cladding of the fibre as mentioned in Chaps. 2 and 3, so the light is interacting with the medium on the surface of the fibre. LPG or TFBG optical fibres may be used for exciting cladding modes. Tapered optical fibres can also be used, but they generally require etching of the fibre cladding.

Due to the significant attenuation of mid and far-infrared in optical fibres and the weak coupling into the cladding it is required to rather functionalize the surface of the fibres with sensor materials absorbing specific gases. The gas binds to the sensor material on the fibre, changes its refractive index, which causes a change in the spectrum of the light either in intensity or as wavelength shifts. Low attenuation can be achieved by using speciality optical fibres manufactured with other materials.

The use of side-faced and end-faced optical fibre designs have great potential for Raman scattering sensors due to the possibility for manipulating the field on the fibre side-face enhancing the Raman scattering. An example of this is SPR, where a sensing material (e.g. palladium for hydrogen detection) causes changes to another material of different permittivity (usually gold or silver), causing a resonant oscillation of conductive electrons at the interface between the two materials. This oscillation resonance can be manipulated by the hydrogenation state of the palladium and can be probed optically.

Raman active molecules may be detected with side-face sensing optical fibres by using either VIS or IR broadband light source. Using the VIS domain has many advantages such as increased scattering since the strength of Raman scattering is inversely proportional to the fourth power of the wavelength. Many wideband sources and detectors are also available and inexpensive. The weak Raman scattering intensity is a significant disadvantage and long measurement times, or high laser power is required to obtain a sufficient signal.

The Raman scattering can also be enhanced by increasing the strength of the light, electromagnetic field interacting with the medium on the fibre side-face. By structuring noble metal or silicon materials on the fibre surface, highly localised and enhanced fields can be obtained. Surface layers may consist of nanometre layer of gold, gold nanoparticles or periodic gratings. The field enhancement amplifies the incident light interaction with the Raman active molecules as well as emitted Raman scattering. The increase of Raman scattering is proportional to the fourth power of the electric field enhancement (i.e. a doubling of the electric field on the fibre side due to structured noble metal layer will result in a 16-fold increase in Raman scattered photons).

In principle, Raman scattering has excellent potential for significantly increasing specificity and sensitivity for measuring gas compositions. However, due to the weak signal obtained with Raman scattering sensors, it is to date less used in industry than IR absorbing sensors. With IR sensors, several methods can be applied to overcome non-specificity due to overlapping absorbing bands or background noise from water. Conventional statistical methods such as partial least square regression or principal component analysis can be applied to analyse complex gas composition.

## Thin Film-Based Optical Fibre Hydrogen Sensors

### Measurement Principles

The two main trends for thin-film based OF $H_2$ sensors are based on two materials. One is palladium-based films, and the other is tungsten oxide-based films in either side-faced or end-faced configurations.

#### Palladium-Based Thin Films

Palladium readily reacts with $H_2$ naturally. When molecular $H_2$ is near palladium, it becomes dissociated into atomic hydrogen. The hydrogen atoms can then diffuse into the palladium, rapidly converting the palladium into the reversible palladium hydride. The crystallographic change in the palladium leads to an incremental increase in the volume of the film and a reduction in the dielectric constant. This leads to changes in the optical signals through an OF related to intensity, wavelength and phase (this depends on the configuration of the sensor). Therefore, these changes can be probed and monitored in response to the $H_2$ concentration near the OF [2–4]. Unfortunately, due to the cyclic absorption and desorption of hydrogen, palladium films are susceptible to cracking, blistering and delamination from multiple expansion and contraction cycles. This embrittlement of the film reduces the stability and sensitivity of the sensor significantly.

#### Tungsten Oxide-Based Thin Films

When tungsten oxide is exposed to $H_2$, its reflectance, transmittance, absorption and refractive index are changed. This makes tungsten oxide an ideal material for optical detection of $H_2$ [5]. Tungsten oxide does not selectively react with $H_2$, as many other gases such as $H_2S$ and acetylene also react with tungsten oxide. Additionally, as it is an oxide, $O_2$ can severely reduce the ability of tungsten oxide to react with $H_2$

through competition. The interaction of tungsten oxide with $H_2$ is also an exothermic reaction, which is not ideal for $H_2$ applications.

## Measurement Methods

Since palladium is sensitive to $H_2$ and can be applied in thin films onto OFs, it is a good candidate for an OF-based $H_2$ sensor. This means that the sensor can be isolated, on-line and measured from a distance. The means of measurement can be either by transmission or reflection of an optical signal through an OF.

### Intensity-Based Hydrogen Sensors

Intensity-based $H_2$ sensors function on the principle that the concentration of $H_2$ will directly modulate the light intensity through the OF. This can be classified as either evanescent field sensors or micro-mirror sensors. Evanescent field sensors are generally the most commonly used type of sensor for $H_2$ applications. The sensing region is in the form of etched, tapered or side-polished OFs with the thin films coated on the core-exposed OF [2, 4].

When the OF cladding is removed, the thin film applied will act as the cladding. This allows the film to modify the evanescent field, influencing the transmission of light through the OF. In the case of palladium as the thin film, this transmission of light is directly modulated by the presence of $H_2$, meaning the transmission intensity is linked to the $H_2$ concentration.

Micro-mirror sensors are another way of producing an intensity-based $H_2$ sensor. These consist of the thin film being applied to the end of an OF to form a mirror [2, 4]. The sensing mechanism relies on the Fresnel reflection phenomenon. Moreover, when $H_2$ interacts with the thin film, the refractive index changes causing the reflection of light to change. This changes the transmission of light through the fibre relative to the refractive index of the thin film. In the case of palladium, this results in a light transmission that is linked to the $H_2$ concentration.

Both micro-mirror and evanescent $H_2$ sensors have straight forward structures and do not demand complicated optical devices. This reduces the cost and makes them more suitable for industrial distribution. Despite this, they are both susceptible to fluctuations in the light source and losses along the OF. Furthermore, the mechanical stability of such sensors is low due to the requirement for removing the OF cladding.

### Fibre Grating-Based Hydrogen Sensors

As described in Chap. 2, fibre gratings are periodic structures that can modulate the refractive index of an OF. This is achieved by inscribing a constant modulation period within the core of the OF. The difference in the modulation period can be grouped into those with a period gap $<1$ μm (FBG), of with a period in the range of 10–100 μm (LPG). For FBG, deception of $H_2$ relies on the variation of strain induced by the palladium on the surface of the fibre, whereas, for LPG, the $H_2$ detection relies on the coupling between the cladding modes (or thin-film) and the evanescent waves.

**Interferometer-Based Hydrogen Sensors**

As discussed in Chap. 2, interferometer-based sensors work on the principle that two light with identical frequencies, a constant phase difference and the same transmission direction will interfere with each other at a specific meeting point. With regards to a $H_2$ sensor, it is assumed that the $H_2$ reacts with a sensitive film (palladium), which induces a stress on the interferometer device. This stress can then affect the phase difference between the two lights, causing a shift in the interference spectrum. The three main options for the structural style of the interferometer for $H_2$ sensing are a Mach–Zehnder and Fabry–Perot interferometers. Mach–Zehnder interferometers have two light inputs with different optical path lengths, causing an out-of-phase interference when the two light paths meet. Fabry–Perot interferometers rely on the reflection of the light back down the OF. To detect $H_2$, palladium is used as the reflective film, causing a shift in the reflective resonant spectrum.

**SPR-Based Hydrogen Sensors**

Surface plasmons are the oscillations of electrons in metal that are stimulated by transverse magnetic light. When the propagation constant of the light matches the propagation constant of the surface plasmon wave, energy is transferred from the incident light wave to the surface plasmon wave. This resonance condition of the surface plasmons is highly sensitive to the dielectric constant of materials in contact with the metal, the angle of incidence and the wavelength of the light. When there is a fixed angle of incidence of light, the resonance occurs at a fixed wavelength termed the resonant wavelength. When changes occur to the dielectric constant of the material touching the metal, the resonance parameter of the surface plasmons changes. The metals gold, silver and copper, are ideal thin layers that support surface plasmons, and the addition of a thin layer of palladium above metal allows modification of the surface plasmon on the support metal that can be observed spectroscopically (Fig. 4.1).

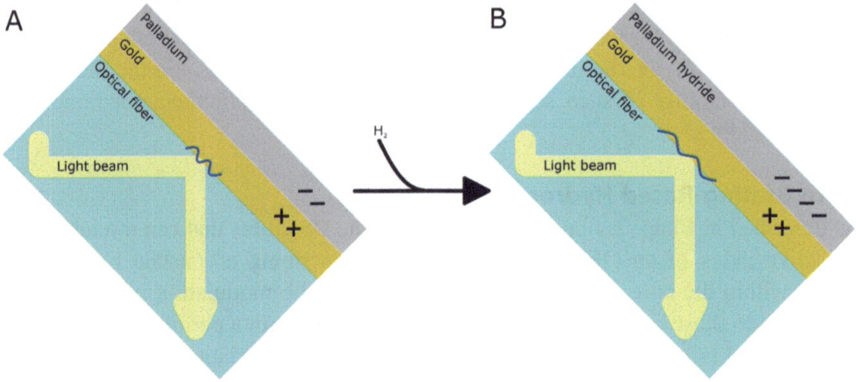


**Fig. 4.1** Optical sensor technologies using surface plasmon resonance

**Acknowledgements** The authors are grateful to the ENERSENSE programme and NTNU Team Hydrogen at the Norwegian University of Science and Technology (NTNU) for supporting and helping on this book project.

## References

1. Hosseini SE, Wahid MA (2016) Hydrogen production from renewable and sustainable energy resources: promising green energy carrier for clean development. Renew Sust Energ Rev 57:850–866
2. Zhang Y, Peng H, Qian X, Zhang Y, An G, Zhao Y (2017) Recent advancements in optical fiber hydrogen sensors. Sensors Actuators B Chem (Internet) 244:393–416. https://www.sciencedirect.com/science/article/pii/S0925400517300047
3. Javahiraly N (2015) Review on hydrogen leak detection: comparison between fiber optic sensors based on different designs with palladium. Opt Eng 54(3):30901
4. Lamb JJ, Bernard O, Sarker S, Lien KM, Hjelme DR (2019) Perspectives of surface plasmon resonance sensors for optimised biogas methanation. Eng Life Sci 19(11):759–769
5. Zheng H, Ou JZ, Strano MS, Kaner RB, Mitchell A, Kalantar-zadeh K (2011) Nanostructured tungsten oxide–properties, synthesis, and applications. Adv Funct Mater 21(12):2175–2196

# 5 Sensor Fusion

Harald I. Muri, Markus Wahl, Jacob J. Lamb, Rolf K. Snilsberg, and Dag R. Hjelme

The fusion of sensors or data is today often used for increasing precision in navigation, position and location of mobile objects in the shipping industry, GPS systems, and smartphones. This is due in part that some information may not be reliable when using the sensor data sources individually. By fusing several independent sensors containing individual information, their combined data may represent precise and usable information or distributions.

Furthermore, sensor fusion can be an improvement of single sensor measurements with sensory deprivation, limited spatial coverage, limited temporal coverage, imprecision or uncertainties being reduced. Therefore, sensor fusion can be designed in competitive, complementary or cooperative configurations. This chapter will mainly focus on complementary fusion where sensor fusion can be used to increase precision or reduce uncertainties by using multiple OF sensors for monitoring different parameters.

H. I. Muri · M. Wahl · R. K. Snilsberg · D. R. Hjelme
Department of Electronic Systems, ENERSENSE, Norwegian University of Science and Technology, Trondheim, Norway
e-mail: harald.muri@ntnu.no; markus.s.wahl@ntnu.no; rolf.k.snilsberg@ntnu.no; dag.hjelme@ntnu.no

J. J. Lamb (✉)
Department of Electronic Systems, ENERSENSE, Norwegian University of Science and Technology, Trondheim, Norway

Department of Energy and Process Engineering, ENERSENSE, Norwegian University of Science and Technology, Trondheim, Norway
e-mail: jacob.j.lamb@ntnu.no



J. J. Lamb, B. G. Pollet (eds.), *Micro-Optics and Energy*,
https://doi.org/10.1007/978-3-030-43676-6_5

## Principle of Sensor Fusion

The terminology of fusing sensory data has often been defined as data fusion, sensor fusion, information fusion or multisensor integration. The terminologies are often used for online sensors, offline sensors or intelligence data used in combination with offline and online sensors. Tools and methods are used to combine a set of intelligence data or sensor data from different sources, so the resulting information is better than if the sources were used individually [1]. The resulting information may be obtained from independent heterogeneous or homogeneous sensor sources. Some limitations of using single sensors may include:

**Limited Temporal Coverage** Single sensors may suffer from long response times due to the processing of data such as frequency filters, curve fitting and statistical analysis. Long response times may also be due to the collection of measurements to obtain a sufficiently high signal.

**Limited Spatial Coverage** Single sensors may only observe a limited section of a process with spatial differences. The single sensor may only then detect the likelihood of this state that can largely differ from the mean average. The use of several sensors may resemble the mean average state of a process with spatial differences.

**Sensory Deprivation** The failure of a sensor causes a loss in the observation of a process. By employing sensor fusion, it is still possible to observe a process with sensor failures.

Evolution in nature has already implemented multisensor integration of acoustic sensing (hearing), light detection (sight), pressure detection (touching) and chemical detection (taste and smell) to increase the precision in measuring the observed environment. For humans, our precise perception of the environment is highly dependent on using all our five senses to estimate the state of the surrounding nature. The state (temperature) of water may not be accurately estimated only by using either of the five senses individually. Evaporation of water can be observed with our eyes to estimate water temperature. Our touch sensors can feel the temperature of the environment (cold, hot) to relate observed water vapour with external temperature.

When combining sensors, the total information may be represented as follows. In cold environments, the difference in temperature between hot water and the external environment is high, and water may also evaporate at temperatures far from boiling point. With human hearing, water boiling can be sensed acoustically without detecting it with sight. The resulting information is perceived as high-temperature water, which we can process to decide not to make physical contact with the water. This exemplifies how the combination of information from different sensors are increasing the precision and reducing uncertainty in the state estimation of a parameter.

## Sensor Fusion Possibilities

The sensor fusion possibilities may be systemised into three categories that summarise their configuration.

**Cooperative Configuration** The sensor fusion may be cooperative configured when two independent sensors monitor the same parameters to derive more information about a process than the information obtained when using a single sensor. Stereoscopic vision is a cooperative sensor fusion configuration where several images are obtained at different viewpoints. The different images at different viewpoints are then used to resemble a three-dimensional image of the observed object [2].

**Competitive Configuration** The sensor configuration can be competitively configured if the sensors monitor independent parameters of the same property. This way, the sensors with the largest signal-to-noise ratio, or the sensors that have a response accordingly to a statistical model, will be included in the parameter computations. Competitive systems can often be combined with the cooperative configuration.

**Complementary Configuration** A complementary sensor fusion configuration applies sensors that either monitor two different environments or monitors the same environments with two different and independent sensor systems with different signal and noise functions. By using multiple cameras to observe separate parts of a room or using an electronic and an optic sensor in combination to monitor temperature can be categorised as a complementary sensor fusion configuration. This configuration can offer an increase in monitoring precision of a process that would not be possible with single independent sensors.

## Data Handling

The simplest form of sensor fusion can be obtained by using two different sensor parameters and by performing an averaging process, or to choose the maximum value of energy and transfer it to the result. The signal to noise ratio may be increased by fusing two sensor responses $A\times1$ and $A\times2$. The wavelet transforms contain a signal, $a$ and $b$, and an error or noise as $c$ and $d$. The sensor fusion has obtained an average of the signal, and error $A\times1$ and $A\times2$. The result increased signal to noise ratio from $a/b$ to $(a + b)/(c + d)$. For the maximum value method signal to noise is also increased.

The Kalman filter method may increase precision for systems such as the two-sensor fusion example above [3]. Instead of averaging over the signal and noise, each sensor parameter measurement is weighted differently based on its variance $\sigma^2$. The sensors with minimal covariance are weighted more than the sensors with higher covariances. This leads to filtering out noise from using the multiple sensors with independent parameter measurements. For a simple sensor fusion using two sensors performing one measurement only, the Kalman filter can be used to compute the best estimate of the state as:

$$\widehat{x} = z_1 + K(z_2 - z_1) \tag{5.1}$$

$$\sigma^2 = (1 - K)\sigma_1^2 \tag{5.2}$$

$$K = \frac{\sigma_1^2}{\sigma_1^2 + \sigma_2^2} \tag{5.3}$$

where $z_1$ and $z_2$ are the two sensor measurements, $K$ is Kalmar gain, and $\sigma_1^2$ and $\sigma_2^2$ are the gaussian noise of sensor measurement $z_1$ and $z_2$, respectively. The variances from the two sensors are used to weigh the measurement to find the best estimate of the state.

Note that the statements above are a simplification of using a Kalman filter for sensor fusion. The filter may be extended for non-Gaussian noise and for time dependence. In most situations, it is not only a low signal to noise that is the issue but also drift in measurement due to the time dependence of the sensor monitoring the state of a system. Some extensions of Kalman filter methods are:

**Non-linear Systems** Since linear modelling is not always possible, extended Kalman filter methods have been derived that use non-linear stochastic difference equations to perform the system modelling [4].

**Estimation of System Parameters** Statistical parameters are not always prior known and are not constant over time. Therefore, a version of the Kalman filter method has been produced to estimate statistical parameters [5].

**Least Mean Square Optimisation Alternatives** The use of the least mean square approach minimises the error in the Kalman filter method. Other alternatives may be better suited for specific applications, such as the $H_\infty$ norm [6].

Other sensor fusion algorithms of importance are also support vector machine, Bayesian inference technique and fuzzy logic.

**Bayesian Inference Technique** Multiple sensor parameters can be used to find the probability of a process to be described in a specific state, such as a rock blocking the road. The process is described either being in a state of a clear or blocked road. The multiple sensors are providing information about whether there the road is blocked or not and by using the inference technique, the highest probability is found for either of these states. This sensor fusion is increasing the completeness in estimating an observed process.

**Fuzzy Logic** Fuzzy logic is a method where the uncertainty in multisensor fusion can be categorised in the inference processes. The inconsistent information from the sensor parameters is "truth" valued with a number between 0 and 1. Based on the process observed, the most trustworthy sensor parameters are valued as significant depending on the rule applied. The rules added to the model may also be weighted differently statically or dynamically.

**Support Vector Machine** Multiple sensor parameters can be combined with support vector machine algorithms that are supervised, learning models. Classification and regression analysis about a process is performed in advance and serves as a set of training examples. The SVM is optimising a hyperplane based on the input sensor parameters and minimises structural error or maximising margins. This sensor fusion method is increasing the completeness of an observed process by including several sensors.

Sensor fusion allows physical limitations to be overcome in sensor systems. There are a host of different fusion techniques, some of which have been briefly covered here. Although many methods have been left out, Kalman Filter and Bayesian reasoning methods are generally used most frequently when fusing sensors.

**Acknowledgements** The authors are grateful to the ENERSENSE programme and NTNU Team Hydrogen at the Norwegian University of Science and Technology (NTNU) for supporting and helping on this book project.

## References

1. Sarma VVS, Raju S (1991) Multisensor data fusion and decision support for airborne target identification. IEEE Trans Syst Man Cybern 21(5):1224–1230
2. Hoover A, Olsen BD (2000) Sensor network perception for mobile robotics. In: Proceedings 2000 ICRA millennium conference IEEE international conference on robotics and automation symposia proceedings (Cat No 00CH37065). IEEE, pp 342–347
3. Lamb JJ, Bernard O, Sarker S, Lien KM, Hjelme DR (2019) Perspectives of optical colourimetric sensors for anaerobic digestion. Renew Sustain Energy Rev 111:87–96
4. Bishop G, Welch G (2001) An introduction to the Kalman filter. Proc SIGGRAPH, Course 8 (27599–23175):41
5. Åström KJ, Wittenmark B (2013) Computer-controlled systems: theory and design. Courier Corporation, North Chelmsford
6. Shaked U, Theodor YH (1992) sub infinity/-optimal estimation: a tutorial. In: Proceedings of the 31st IEEE conference on decision and control. IEEE, pp 2278–2286

# Part III

# Energy Production and Storage

# Hydrogen Fuel Cells and Water Electrolysers

6

Jacob J. Lamb, Odne S. Burheim, and Bruno G. Pollet

## Introduction

Energy storage in the form of $H_2$ is in many cases considered to be the best means to store energy coming from intermittent (e.g. wind and solar) renewable energy sources. With localised capacities for renewable energy sources proliferating, a storage system that can match the production rate is urgently required. Proton exchange membrane (PEM) fuel cells and electrolysers may provide the basis for a sustainable $H_2$ production solution that is suited for being coupled to intermittent renewable energy sources.

The interest in PEM technology has been revitalised in recent years with the increasing interest in producing green energy; however, there are many challenges to be explored to improve the technology for wider distribution. Furthermore, a high level of understanding of the internal temperature distribution within a PEM system will allow computational modelling to increase in accuracy, aiding the conceptual development of PEM systems.

With regard to PEM systems, more than 94,000 fuel cell studies have been published, and almost all water electrolysis methods are also used for fuel cell research, suggesting that water electrolysis research has investigated methods used in fuel cell research [1, 2]. This may be related to fuel cell research encouraging the development of water electrolysis technologies. Therefore, for the context of

J. J. Lamb (✉)
Department of Electronic Systems, ENERSENSE, Norwegian University of Science and Technology, Trondheim, Norway

Department of Energy and Process Engineering, ENERSENSE, Norwegian University of Science and Technology, Trondheim, Norway
e-mail: jacob.j.lamb@ntnu.no

O. S. Burheim · B. G. Pollet
Department of Energy and Process Engineering, ENERSENSE and NTNU Team Hydrogen, Norwegian University of Science and Technology, Trondheim, Norway
e-mail: odne.s.burheim@ntnu.no; bruno.g.pollet@ntnu.no





J. J. Lamb, B. G. Pollet (eds.), *Micro-Optics and Energy*,
https://doi.org/10.1007/978-3-030-43676-6_6

developing micro-optical sensors for determining temperature distribution in PEM systems, the fuel cell and electrolyser alternatives are virtually identical. Since the PEM fuel cell has developed further than the PEM electrolyser, after a brief description of PEM electrolysers, this chapter will focus more on the fuel cell design.

## Hydrogen Production

### Traditional Production

The use of $H_2$ as an economically viable energy source has been held back initially by the fact that molecular $H_2$ (unlike fossil fuels) does not readily exist in large scales in nature. Therefore, to obtain $H_2$, energy input is required. Most commonly, $H_2$ is produced by steam reforming of natural gas or other fossil fuels, such as propane, gasoline, diesel, methanol or ethanol [3–5]. This is achieved in what is known as a reformer, where high temperature (e.g. 700–1000 °C) water vapour reacts with fossil fuel in the presence of a catalyst generally constructed of nickel [3–5]. Although widely used, the steam reforming method of $H_2$ production generally produces low purity $H_2$ with many carbon-containing species (e.g. CO). Furthermore, by using fossil fuels to produce $H_2$, steam reforming does not relieve the dependencies on fossil fuels; therefore, it does not contribute to the energy matrix as a clean, renewable energy storage solution.

Water electrolysis is a method for production of $H_2$ through the electrochemical conversion of water into its two constituents, $H_2$ and $O_2$ [6]. This can yield high-purity $H_2$ due to the lack of complex chemistry occurring.

### Electrochemical Production

Water electrolysis is currently the most prominent primary method for sheer $H_2$ manufacturing by electrochemical means, and it is expected that its importance will expand rapidly shortly. Electrolysis of water is based on the movement of electrons supported by an external circuit. The main electrochemical $H_2$ manufacturing techniques are alkaline, proton exchange membrane (PEM) and solid oxide (SO) electrolysers. The input into the electrolyser will be dynamic due to the periodic availability of renewable energy from solar and wind sources. Due to their flexible operation, PEM electrolyser technology is preferential in a renewable energy sense over alkaline and SO techniques.

The main advantages of PEM electrolysers are as follows:

- There is a possibility of operating the cells at a high current density.
- High efficiencies can be obtained, even at high current densities.
- High-purity gases are produced as a result of the high purity water used.
- They have a high dynamic range.

The main drawbacks of PEM electrolysers are as follows:

- Capital expenses are still very high.
- Need for high-purity water.
- Large-scale systems are not yet developed.

## Turning Hydrogen into Electricity

There are many different types of fuel cells, and PEM fuel cells have been developed significantly more than their electrolyser alternatives. Moreover, PEM fuel cells have been developed the most and implemented more due to their capability for generating higher power densities, higher efficiencies, low weight, compact size, lower cost, lower operation temperatures and faster start-up times when compared to the other types of FCs [7–10].

The PEM is the most essential part of a PEMFC. It must be impermeable to gases and not conduct electrical current, but it must allow passage of protons. Therefore, the membrane acts somewhat like an electrolyte and is placed between two porous, electrically conductive electrodes. These electrodes typically are constructed from carbon-based cloth or paper with platinum catalyst deposited on it. Together, these components make up the membrane electrode assembly (MEA). At both sides of the MEA is where the bipolar plates (BPP) are found. They provide the physical structure of the FC stack, conducting current between single cells, provide control of the temperature via cooling channels, and achieve a uniform reactant gas distribution via their gas channel flow patterns. To avoid corrosion and surface contact resistance, BPPs are generally constructed using graphite composites. Backing layers constructed of polytetrafluoroethylene (PTFE)-coated carbon fibres are known as the gas diffusion layers (GDL). These are used to diffuse the reactant gases and electrons uniformly while limiting the accumulation of liquid water in the MEA. Between the PEM and the GDL is a catalyst layer with catalytic nanoparticles (typically a thin layer of platinum or a platinum-based alloy spread on particles of carbon that are embedded within the polymer electrode). The electrochemical reactions are catalysed in this layer.

The operation principle of the PEM fuel cell is that $H_2$ is supplied to the anode and $O_2$ to the cathode. This is achieved through gas flow channels, resulting in the generation of protons at the anode and current through an external circuit. Moreover, at the anode, $H_2$ is oxidised into protons and electrons. The protons are transported through the PEM, and the electrons through the electrodes, followed by the current collectors, and then through the external electrical circuit to the cathode. The electrons combine with protons and $O_2$ at the cathode electrode to produce water and heat. The overall result of these reactions is the direct current (DC) generation for external usage.

## Effects of Temperature and Humidity Within PEMFCs

In order for PEM fuel cells to become a commercial energy production system, several technological challenges regarding thermal management must be resolved [7, 8, 11–14]. This management includes heat removal from the PEM system generated during regular operation to prevent overheating. The heat can be dissipated using heat exchanging loops or via air convection [15].

The general efficiency of a PEM fuel cell is around 50%, meaning that almost half of the $H_2$ energy is converted into heat. Long-term stability of a PEM fuel cell is, therefore, reliant on thermal management being addressed [16–18]. High-resolution analysis of the temperature distribution of the amount of heat generated within the PEM fuel cells will be required to improve their efficiency [19–21]. Ideally, each cell within a PEM fuel cell stack should operate with a uniform temperature distribution; however, variations in temperature arise from the inherent design of the manifolds of the stack, the position of single cells within the stack and the method chosen for cooling [22]. Moreover, the maximum cell voltage (theoretical) that a PEM fuel cell can operate is determined by the temperature of the fuel cell [23]; therefore, it can be assumed that an increase in the cell temperature would result in a lower cell voltage in theory [24].

Temperature distributions in PEM fuel cells can be explained as a function of many transport phenomena within the PEM fuel cell, including multicomponent gas transport, charge transport, and two-phase flow and heat transfer, all of which take part in different components of the PEM fuel cell [11]. In order to avoid overheating, the excess heat generated must be removed from the PEM fuel cell. In most PEM fuel cells; the heat is removed by natural convection and radiation. Part of this heat removal is achieved by unused reactants leaving the PEM fuel cell stack, while active cooling is put in place to avoid overheating [25]. In lower temperature PEM fuel cells (operating range of 60–80 °C), a higher temperature can rapidly increase the rate of PEM and catalyst degradation, therefore reducing the stack performance [25–30].

Conversely, low temperatures are also not favourable, as they may cause electrode flooding due to a lower water saturation pressure required at a lower temperature, which can cause significant problems about water management within the PEM fuel cell [31]. Furthermore, if there is too much heat energy removed from the PEM fuel cell by active cooling, the kinetics of the chemical reactions within the cell is negatively affected. This can cause the partial pressure of water vapour to increase to a level higher than the saturation pressure, which leads to condensation within the PEM fuel cell, lowering the performance significantly.

It can also become disastrous for the PEM fuel cell if the temperature within becomes is too high. This can occur if the active cooling is not adequate to remove excess heat energy, and the temperature can rise above the allowable operating temperature, which can cause dehydration of the cell, reducing the protonic conductivity of the cell, reducing the stack performance and stability [31, 32]. Localised heat increases within the PEM fuel cell due to an insufficiently designed cooling system can lead to the degradation of the components within the PEM fuel cell, significantly reducing the reliability.

The primary purpose of temperature management in PEM fuel cells is to ensure that the PEM fuel cell stacks operate within what is deemed to be a reliable temperature range and maintaining a uniform distribution of heat [33]. This heat energy can also be recycled, allowing for its effective use for other requirements, improving the efficiency of the PEM fuel cell [34].

Within a PEM fuel cell, an increase in operating temperature up to 80 °C can affect the stack performance positively; however, once the temperature reached 85 °C, the stack performance starts to drop rapidly. This is because the conductivity of the polymer membrane is reduced as there is a lower water content [9, 33]. Therefore, it is critical with regard to the performance of the PEM fuel cell that the temperature is maintained at an ideal operating temperature between 60 and 80 °C. This becomes imperative when the PEM fuel cell is running at high current densities, as heat is generated at a higher rate and needs to be quickly removed.

## Distribution of Temperature and Humidity Within PEMFCs

The heat generated within PEM fuel cells generally originates from the cathode side of the catalyst layers, but the membrane and the electrically conductive parts of the PEM fuel cell can also generate heat. Cross-sectional analysis of a PEM fuel cell can show the gradient distribution of heat. The generation of heat in a PEM fuel cell is the result of four factors:

1. The irreversibility of the electrochemical reactions
2. The heat of the reactions
3. The ohmic resistances
4. The condensation of water

The difference between the total chemical energy of the reactants and the maximum of available work is known as entropic heat, in accordance with the Second Law of Thermal Dynamics. The entropic heat is representative of the change in entropy of the electrochemical reactions and must be either supplied to or subtracted from the electrode regions of the PEM fuel cell. What is known as the irreversible heat results from the electrochemical reactions irreversibility within the PEM fuel cell and can be a significant heat generator. Ohmic heating is generated by the inherent resistance of the electrolyte, electrodes, the bipolar plates and the current collecting components within the PEM fuel cell [32, 35, 36]. In addition, the entropic heat associated with the condensation and evaporation of water within the gas diffusion layers must be taken into account.

Heat generation within PEM fuel cells is driven by the electrode thermal resistance properties, the electrochemical rates of reactions, the uniformity of the reactions, the humidity of the reactants and the permeability of the PEM/GDL. The rate at which the heat is generated is proportional to decreases in cell voltages and increases in current densities. These electrical properties are indirectly proportional, because as the current density increases, the cell voltage decreases, and this significantly increases the rate of heat generation within the PEM fuel cell.

The distribution of the heat generated within the PEM fuel cell stack is uneven. Furthermore, the entropy variation imbalance and electrode losses, the exothermic electrochemical reactions that take place at the cathode side significantly influence the heat generation within the PEM fuel cell [21]. Heat conduction is the primary mode of thermal transport through the PEM, while the transfer in the catalyst and gas diffusion layer is significantly influenced by conduction and convection of heat [37]. The convective heat flux direction is the same as the flow direction within the PEM fuel cell; however, the conductive heat flux is in the opposite direction as it is determined by the temperature gradient [31, 37]. The rate of local heat generation affects the durability and performance of the PEM fuel cells, is not straight forward to measure and varies in both through-plane and in-plane directions.

A significant disparity in the understanding of the heat distribution in the through-plane still exists. This is because the heat generation is usually non-uniform within the in-plane direction because of the interrelationship of the water content, local current density, reactant concentration and temperature. The non-uniform distribution of the local heat generation and temperature significantly complicates the cooling of PEM fuel cell stacks [19].

The removal of heat within the PEM fuel cell is achieved using some form of the cooling system or transferred across the faces of the PEM fuel cell stack by conduction and convection. The rate of heat removal is dictated by each component (i.e. PEM, GDL, catalyst layer and BPP) thermal properties. The amount of chemical energy that can be converted into electricity is determined partially by the state of the water produced by the fuel cell (either liquid or gas). The absorption and release of latent heat, water and heat transport are bound to the evaporation and condensation processes of water. These occur in combination with each other due to what is known as a "heat pipe effect." The vapour pressure saturation point is heavily dependent on local temperatures [38]. The significant contributors to heat generation in the PEM fuel cell are the electrochemical reactions, with the most severe water and heat management problems occurring at the cathode catalyst layer due to its large heat generation. Evaporation of the liquid phase is stimulated by the small pores within the catalyst layers, allowing the flow of water out of the system in the vapour phase, enabling the chemical reactants to diffuse towards their site of reaction. The cathode catalyst layers generated heat can also be conducted through the GDL, BPP and subsequently removed by active cooling. Therefore, the thermal conductivity of the cathode catalyst layer is of significant interest in developing temperature management further [39].

When the high temperature-vaporised water passes through the MPL and GDL, it is slightly cooled, and some water vapour is condensed into liquid water. This improves the heat transfer as latent heat is released during water condensation. Therefore, the thermal gradients within the GDL have a significant effect on water transport and condensation [40].

The gas diffusion layers used for PEM fuel cell construction are highly directionally dependent when it comes to heat transport properties. Moreover, they have different properties for through-plane and in-plane transport directions. When designing the GDL, the heat transfer properties are fundamental for the efficient operation of a PEM fuel cell. The effective thermal conductivity and thermal contact resistance between the GDL and all neighbouring layers must be analysed thoroughly during the design phase [41, 42].

Detailed knowledge of the in situ distribution of temperature within the PEM fuel cell is fundamental for the improvement and optimisation of PEM fuel cell performance and durability. Although through-plane thermal conductivities can be used for predicting temperature distribution, in-plane conductivities are crucial for improving discrepancies between modelled and experimental data [43]. There are two additional fundamental parameters other than the GDL thermal conductivity that significantly influence the PEM fuel cells thermal behaviour. The relative humidity of the feed gas is one of these fundamental parameters, as this has a significant effect on the overall membrane hydration. When the feed gas is fully humidified, the thermal effect is reduced as the membrane maintains full hydration. The other critical parameter is the cell voltage, as this directly controls the current output and therefore the heat generation rate [44].

## Research Needs and Measurement Challenges

Managing heat in PEM fuel cells is traditionally secondary in comparison to performance. Despite this, in recent years, it has become a crucial area of improvement for the continuation of performance development [24]. There is a need for more detailed knowledge in the field of PEM fuel cell science and technology development. To understand the challenge, one must understand the geometric levels that a PEM fuel cell is assembled and how temperatures, water and other parameters give an impact.

The core reactor of a PEM fuel cell is the membrane and its anode and catalyst layer (AC and CC). It is in these three layers that the heat is generated. The heat is generated in very thin layers (10–50 μm) and then has to be transported through several thicker layers (200–1000 μm), and this leads to substantial temperature gradients.

A commonly modelled area is indicated with pink and the boundary lines with blue, as shown in Fig. 6.1. The distance between the blue lines here is typically 1000 μm. Using a continuum model with evenly distributed reaction rate and considering different thermal conductivity values [45], one can see that the local temperature variations are up to 14° between the backing reference and the membrane (Fig. 6.2). The research problem is complicated, as the temperature, thermal conductivity and thermal gradients are all within the range of 10–1000 μm. One can see that measuring this temperature requires small sensors.

The understanding of local humidity, temperature and localised gas pressures can be improved by experiments and by modelling. To avoid modelling, which is too empirical and with too many assumptions, high resolution and highly detailed

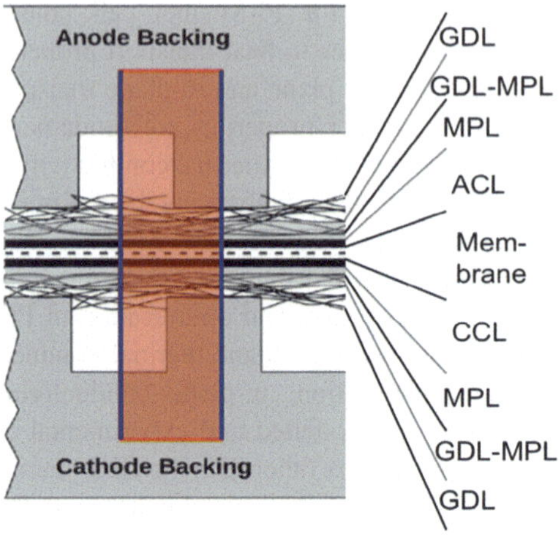


**Fig. 6.1** PEM fuel cell assembly

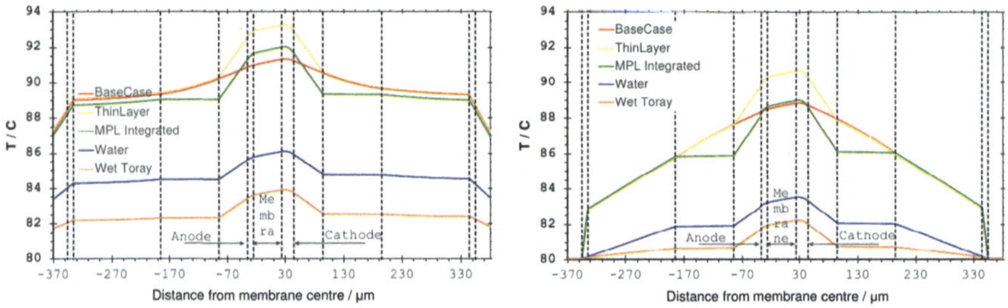


**Fig. 6.2** Left: Land temperature profiles. Right: Channel temperature profiles

experimental results are needed that cover local temperature and relative humidity in different positions within the PEM fuel cell. The experimental challenge is that catalyst layers are 10–20 µm thick and existing sensors are at least twice as large, and that is before adding corrosion protective coatings. When the sensor (with or without their coating) is larger than the region of interest, one cannot know where data are obtained. Detailed measurements down to the precision of 10 µm are required.

Improved knowledge of data point location for improved understanding of heat and mass transport and improved understanding of ageing mechanisms of a PEM fuel cell is essential in this field. Optical-based sensors that can function on other electrochemical devices with regions thinner than 10 µm is a promising route.

## Possibilities for Micro-optical Technologies in PEMFCs

Currently, research into the thermal properties of PEM fuel cells relies on thermocouples that are beyond 100 µm thick when including coatings [45] and the electrical components [46] of the sensor. The challenges with these sensors are twofold. They are susceptible to corrosion-based degradation, and they have a large relative size (the sensor thicknesses influence on the thermal transport within PEM fuel cells is not yet entirely defined). Therefore, there is a requirement for smaller non-electrical sensors to measure temperature within PEM fuel cells.

There is a possibility for micro-optical sensors with small spatial distribution to be designed to undertake fundamental investigation targeting the measurement of temperature within PEM fuel cells. This will allow the research community to have a stronger understanding of the heat distribution within PEM fuel cells. For example, this could improve the understanding of the thermal properties of the PEM fuel cells, allowing the design and manufacture of PEM fuel cell components (e.g. GDL), with significantly improved heat management properties and power performance. The primary requirement of the production of micro-optical sensors that can be used in electrochemical systems is to have an inert sensor that is small enough that it has a negligible effect on the internal components of the PEM fuel cell, which has layers thinner than 10 μm.

**Acknowledgements** The authors are grateful to the ENERSENSE programme and NTNU Team Hydrogen at the Norwegian University of Science and Technology (NTNU) for supporting and helping on this book project.

## References

1. Ogawa T, Takeuchi M, Kajikawa Y (2018) Analysis of trends and emerging technologies in water electrolysis research based on a computational method: a comparison with fuel cell research. Sustainability 10(2):478
2. Ogawa T, Takeuchi M, Kajikawa Y (2018) Comprehensive analysis of trends and emerging technologies in all types of fuel cells based on a computational method. Sustainability 10(2):458
3. Ni M, Leung DYC, Leung MKH (2007) A review on reforming bio-ethanol for hydrogen production. Int J Hydrog Energy 32(15):3238–3247
4. Rostrup-Nielsen JR, Sehested J, Nørskov JK (2002) Hydrogen and synthesis gas by steam-and $CO_2$ reforming. Adv Catal 47:65–139
5. Rostrup-Nielsen JR, Rostrup-Nielsen T (2002) Large-scale hydrogen production. CATTECH 6 (4):150–159
6. Lamb JJ, Pollet BG, Burheim OS (2020) Energy storage. In: Energy-smart buildings design: construction and monitoring of buildings for improved energy efficiency. https://doi.org/10.1088/978-0-7503-3259-0ch6. ISBN 978-0-7503-3257-6
7. Yu S, Jung D (2008) Thermal management strategy for a proton exchange membrane fuel cell system with a large active cell area. Renew Energy 33(12):2540–2548
8. Eckl R, Zehtner W, Leu C, Wagner U (2004) Experimental analysis of water management in a self-humidifying polymer electrolyte fuel cell stack. J Power Sources 138(1–2):137–144
9. Pérez-Page M, Pérez-Herranz V (2011) Effect of the operation and humidification temperatures on the performance of a PEM fuel cell stack on dead-end mode. Int J Electrochem Sci 6:492–505
10. Wang Y, Chen KS, Mishler J, Cho SC, Adroher XC (2011) A review of polymer electrolyte membrane fuel cells: technology, applications, and needs on fundamental research. Appl Energy 88(4):981–1007

11. Kandlikar SG, Lu Z (2009) Fundamental research needs in combined water and thermal management within a proton exchange membrane fuel cell stack under normal and cold-start conditions. J Fuel Cell Sci Technol 6(4):44001
12. Kandlikar SG, Lu Z (2009) Thermal management issues in a PEMFC stack–a brief review of current status. Appl Therm Eng 29(7):1276–1280
13. Pandiyan S, Jayakumar K, Rajalakshmi N, Dhathathreyan KS (2008) Thermal and electrical energy management in a PEMFC stack–an analytical approach. Int J Heat Mass Transf 51 (3–4):469–473
14. Nolan J, Kolodziej J (2010) Modeling of an automotive fuel cell thermal system. J Power Sources 195(15):4743–4752
15. Cao T-F, Lin H, Chen L, He Y-L, Tao W-Q (2013) Numerical investigation of the coupled water and thermal management in PEM fuel cell. Appl Energy 112:1115–1125
16. Yan W-M, Chen C-Y, Mei S-C, Soong C-Y, Chen F (2006) Effects of operating conditions on cell performance of PEM fuel cells with conventional or interdigitated flow field. J Power Sources 162(2):1157–1164
17. Park YH, Caton JA (2008) Development of a PEM stack and performance analysis including the effects of water content in the membrane and cooling method. J Power Sources 179 (2):584–591
18. Baek SM, Yu SH, Nam JH, Kim C-J (2011) A numerical study on uniform cooling of large-scale PEMFCs with different coolant flow field designs. Appl Therm Eng 31(8–9):1427–1434
19. Zhang G, Guo L, Ma L, Liu H (2010) Simultaneous measurement of current and temperature distributions in a proton exchange membrane fuel cell. J Power Sources 195(11):3597–3604
20. Hsiao M-C, Liao S-H, Yen M-Y, Ma C-CM, Lee S-J, Lin Y-F et al (2009) Electrical and thermal conductivities of novel metal mesh hybrid polymer composite bipolar plates for proton exchange membrane fuel cells. In: ASME 2009 7th international conference on fuel cell science, engineering and technology. American Society of Mechanical Engineers, pp 871–878
21. Yu S, Jung D (2010) A study of operation strategy of cooling module with dynamic fuel cell system model for transportation application. Renew Energy 35(11):2525–2532
22. Strahl S, Husar A, Puleston P, Riera J (2014) Performance improvement by temperature control of an open-cathode PEM fuel cell system. Fuel Cells 14(3):466–478
23. Sasmito AP, Birgersson E, Mujumdar AS (2011) Numerical evaluation of various thermal management strategies for polymer electrolyte fuel cell stacks. Int J Hydrog Energy 36 (20):12991–13007
24. Bvumbe TJ, Bujlo P, Tolj I, Mouton K, Swart G, Pasupathi S et al (2016) Review on management, mechanisms and modelling of thermal processes in PEMFC. Hydrog Fuel Cells 1(1):1–20
25. Islam MR, Shabani B, Rosengarten G, Andrews J (2015) The potential of using nanofluids in PEM fuel cell cooling systems: a review. Renew Sust Energ Rev 48:523–539
26. Faghri A, Guo Z (2005) Challenges and opportunities of thermal management issues related to fuel cell technology and modeling. Int J Heat Mass Transf 48(19–20):3891–3920
27. Rodatz P, Büchi F, Onder C, Guzzella L (2004) Operational aspects of a large PEFC stack under practical conditions. J Power Sources 128(2):208–217
28. Lasbet Y, Auvity B, Castelain C, Peerhossaini H (2006) A chaotic heat-exchanger for PEMFC cooling applications. J Power Sources 156(1):114–118
29. Banerjee R, See E, Kandlikar SG (2013) Pressure drop and voltage response of PEMFC operation under transient temperature and loading conditions. ECS Trans 58(1):1601–1611
30. Van den Oosterkamp PF (2006) Critical issues in heat transfer for fuel cell systems. Energy Convers Manag 47(20):3552–3561
31. Hosseinzadeh E, Rokni M, Rabbani A, Mortensen HH (2013) Thermal and water management of low temperature proton exchange membrane fuel cell in fork-lift truck power system. Appl Energy 104:434–444
32. Shahsavari S, Desouza A, Bahrami M, Kjeang E (2012) Thermal analysis of air-cooled PEM fuel cells. Int J Hydrog Energy 37(23):18261–18271

33. Ravishankar S, Prakash KA (2014) Numerical studies on thermal performance of novel cooling plate designs in polymer electrolyte membrane fuel cell stacks. Appl Therm Eng 66 (1–2):239–251
34. Cozzolino R, Cicconardi SP, Galloni E, Minutillo M, Perna A (2011) Theoretical and experimental investigations on thermal management of a PEMFC stack. Int J Hydrog Energy 36 (13):8030–8037
35. Barbir F (2012) PEM fuel cells: theory and practice. Academic, Waltham
36. Yan Q, Toghiani H, Causey H (2006) Steady state and dynamic performance of proton exchange membrane fuel cells (PEMFCs) under various operating conditions and load changes. J Power Sources 161(1):492–502
37. See E, Kandlikar SG (2013) Effect of GDL material on thermal gradients along the reactant flow channels in PEMFCs. ECS Trans 58(1):867–880
38. Ramousse J, Deseure J, Lottin O, Didierjean S, Maillet D (2005) Modelling of heat, mass and charge transfer in a PEMFC single cell. J Power Sources 145(2):416–427
39. Lopez-Sabiron AM, Barroso J, Roda V, Barranco J, Lozano A, Barreras F (2012) Design and development of the cooling system of a 2 kW nominal power open-cathode polymer electrolyte fuel cell stack. Int J Hydrog Energy 37(8):7289–7298
40. Ramousse J, Lottin O, Didierjean S, Maillet D (2009) Heat sources in proton exchange membrane (PEM) fuel cells. J Power Sources 192(2):435–441
41. Wu J, Galli S, Lagana I, Pozio A, Monteleone G, Yuan XZ et al (2009) An air-cooled proton exchange membrane fuel cell with combined oxidant and coolant flow. J Power Sources 188 (1):199–204
42. Karimi G, Li X, Teertstra P (2010) Measurement of through-plane effective thermal conductivity and contact resistance in PEM fuel cell diffusion media. Electrochim Acta 55(5):1619–1625
43. Sadeghi E, Djilali N, Bahrami M (2011) Effective thermal conductivity and thermal contact resistance of gas diffusion layers in proton exchange membrane fuel cells. Part 1: Effect of compressive load. J Power Sources 196(1):246–254
44. Zamel N, Li X, Shen J, Becker J, Wiegmann A (2010) Estimating effective thermal conductivity in carbon paper diffusion media. Chem Eng Sci 65(13):3994–4006
45. Burheim OS, Crymble GA, Bock R, Hussain N, Pasupathi S, Du Plessis A et al (2015) Thermal conductivity in the three layered regions of micro porous layer coated porous transport layers for the PEM fuel cell. Int J Hydrog Energy 40(46):16775–16785
46. Reum M, Freunberger SA, Wokaun A, Büchi FN (2009) Measuring the current distribution with sub-millimeter resolution in PEFCs II. Impact of operating parameters. J Electrochem Soc 156(3):B301–B310

# Ultrasound-Assisted Electrolytic Hydrogen Production

# 7

Md H. Islam, Jacob J. Lamb, Odne S. Burheim, and Bruno G. Pollet

## Introduction

The search for alternative, pollution-free and efficient energy sources has given rise to the concept of the Hydrogen Economy [1]. Power ultrasound is a technology that has been used in novel methods to produce various renewable energy sources recently [2–4]. $H_2$ as an energy carrier is particularly attractive due to its very high specific energy (39.40 kWh/kg) compared to other conventional fossil fuels, as shown in Table 7.1. The production of electrolytic $H_2$ from water in the presence of ultrasound is seen as a promising method to produce clean $H_2$ [5, 6]. In this chapter, the production of $H_2$ through the sonoelectrochemical method, along with a brief description of current $H_2$ production methods, is presented.

## Hydrogen Production Methods

Currently, 96% of the global $H_2$ demand is produced by steam reforming of fossil fuels. These processes release vast amounts of greenhouse gases [7]. The conventional $H_2$ production methods are summarised in Fig. 7.1.

M. H. Islam · O. S. Burheim · B. G. Pollet
Department of Energy and Process Engineering, ENERSENSE and NTNU Team Hydrogen, Norwegian University of Science and Technology, Trondheim, Norway
e-mail: md.h.islam@ntnu.no; odne.s.burheim@ntnu.no; bruno.g.pollet@ntnu.no

J. J. Lamb (✉)
Department of Electronic Systems, ENERSENSE, Norwegian University of Science and Technology, Trondheim, Norway

Department of Energy and Process Engineering, ENERSENSE, Norwegian University of Science and Technology, Trondheim, Norway
e-mail: jacob.j.lamb@ntnu.no



J. J. Lamb, B. G. Pollet (eds.), *Micro-Optics and Energy*,
https://doi.org/10.1007/978-3-030-43676-6_7

**Table 7.1** Specific energy and energy density of different fuels

| Fuel types | Specific energy (kWh/kg) | Energy density (kWh/L) |
|---|---|---|
| Diesel | 12.67 | 10.72 |
| Gasoline | 12.89 | 9.5 |
| Ethanol | 8.25 | 6.5 |
| Coke | 7.78 | – |
| Liquified methane | 15.44 | 6.53 |
| $H_2$ (liquid) | 39.40 (HHV), 33.31 (LHV) | 2.79 (HHV), 2.36 (LHV) |
| Compressed $H_2$ (@ 690 bar, and 15 °C) | 39.40 (HHV), 53.31 (LHV) | 1.47 (HHV), 1.25 (LHV) |
| $H_2$ gas | 39.40 (HHV), 53.31 (LHV) | 0.0033 (HHV), 0.0279 (LHV) |

*HHV* High Heating Value, *LHV* Low Heating Value

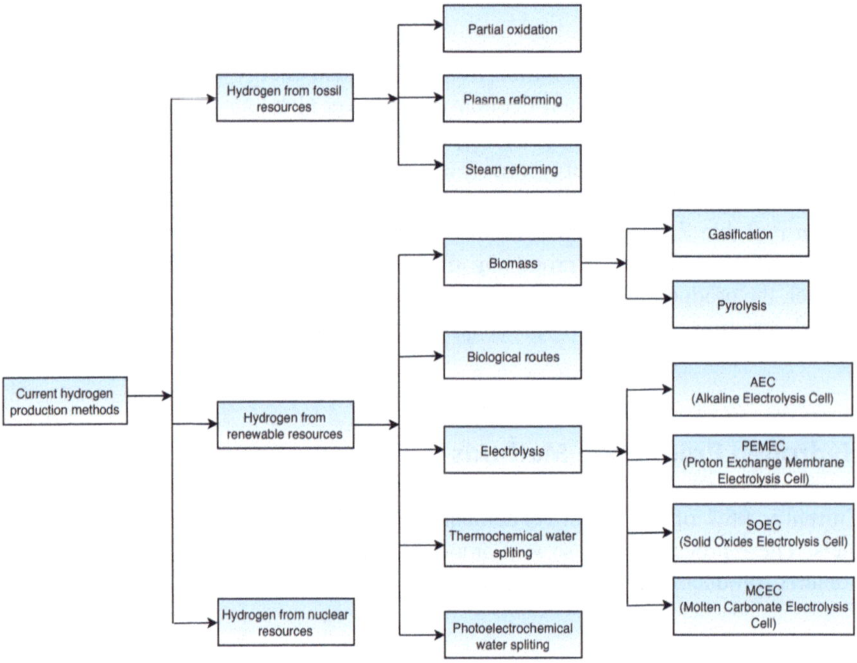


**Fig. 7.1** Current $H_2$ production routes

The production of $H_2$ from natural gas is the most common method. There are three basic methods for $H_2$ production from fossil fuels:

- Steam reforming
- Partial oxidation
- Autothermal reforming.

Steam reforming is highly efficient due to low operating and production costs. Natural gas, lighter hydrocarbons and methanol are the most frequently used materials for steam reforming. The main disadvantage of this process is the high production of $CO_2$ (ca. 7.05 kg of $CO_2$ emitted per kg $H_2$ produced). In partial oxidation, the gasification of the raw material is carried out in the presence of $O_2$ and steam at elevated temperatures (in the range of 1300–1500 °C) and pressure (3–8 MPa). Autothermal reforming (ATR) is a combination of steam reforming and partial oxidation where steam is introduced in the catalytic partial oxidation process. ATR is a more straightforward and less expensive process than steam reforming, and it is more favourable as it does not require external heating [7].

$H_2$ can also be produced from renewable resources avoiding the requirement of reforming fossil fuels. Biomass-based approaches and water electrolysis are the primary sources of renewable $H_2$ production [7]. Gasification of biomass is a promising method where biomass is partially oxidised into a mixture of $H_2$, $CH_4$, $CO_2$, CO and higher hydrocarbons named as "producer gases" [8]. Production of $H_2$ through biological routes offers a wide range of approaches. The primary production routes are direct and indirect biophotolysis, photo-fermentation and dark fermentation [9].

$H_2$ production through water electrolysis is a promising method. Currently, about 4% of total $H_2$ production is obtained through water electrolysis. Based on the electrolytes and working temperatures, the electrolysis of water can be divided into four main categories [1, 10]:

- Alkaline Electrolysis Cell (AEC)
- Proton Exchange Membrane Electrolysis Cell (PEMEC)
- Solid Oxides Electrolysis Cell (SOEC)
- Molten Carbonate Electrolytic Cell (MCEC)

The production efficiency of $H_2$ and the energy consumption of different electrolysers are summarised in Table 7.2, and the chemical reactions and operating temperature range of water electrolysis technologies are shown in Table 7.3.

## Sonoelectrochemical Production of Hydrogen

Sonoelectrochemistry is the pairing of ultrasonic energy with an electrochemical system [6]. The effect of ultrasonic irradiation is not only upon the heterogeneous system (involving the electrode and the electrolyte) but also the homogeneous

**Table 7.2** Summary of electrolyser technologies for hydrogen production [11]

| Electrolyser | Temperature range (°C) | Energy consumption (kWh/kg of $H_2$) | Efficiency (%) |
|---|---|---|---|
| AEC | 60–80 | 53–70 | 56–73 |
| PEMEC | 50–80 | 54–90 | 48–65 |
| SOEC | 600–900 | 35 | 90 |
| MCEC | 600–700 | 35 | 90 |

**Table 7.3** Chemical reactions and operating temperatures of water electrolysis technologies

| Electrolyser technology | Alkaline | Membrane | High temperature |
|---|---|---|---|
| $O_2$ evolution reaction (OER) @ anode | $2OH^- \rightarrow 0.5O_2 + H_2O + 2e^-$ | $H_2O \rightarrow 0.5O_2 + 2H^+ + 2e^-$ | $O^{2-} \rightarrow 0.5O_2 + 2e^-$ |
| $H_2$ evolution reaction (HER) @ cathode | $H_2O + 2e^- \rightarrow H_2 + 2OH^-$ | $2H^+ + 2e^- \rightarrow H_2$ | $H_2O + 2e^- \rightarrow H_2 + O^{2-}$ |
| Temperature range (°C) | 40–90 | 20–100 | 700–1000 |

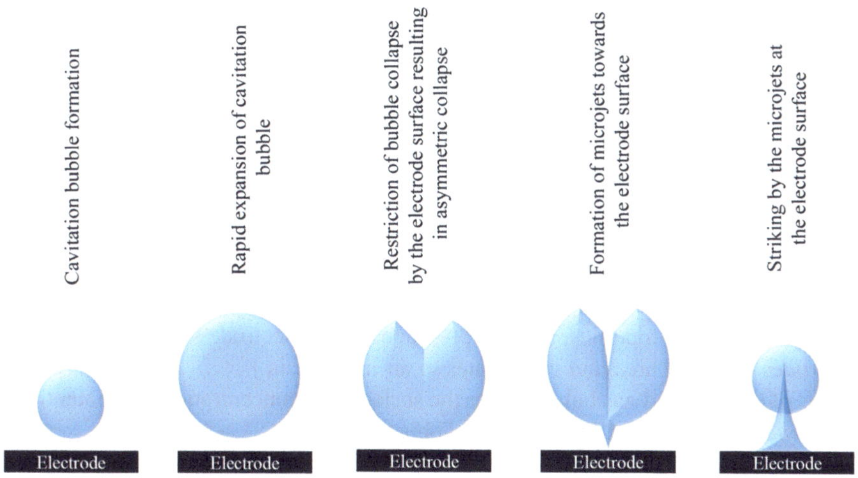


**Fig. 7.2** Bubble collapse at the electrode surface [12]

system that takes place in the bulk solution through extreme conditions (e.g. acoustic cavitation) [6]. The collapsing of acoustic bubbles on a solid surface leads to the formation of microjets being directed towards the surface of the solid material at speeds of up to 200–500 m $s^{-1}$ (Fig. 7.2) [13]. Microstreaming is also caused by the bubble close to the electrode surface [3]. If the surface is an electrode, the combined effect of the microjet and microstreaming significantly promotes mass transport to the electrode surface. Moreover, electrode cleaning and surface activation can also be enhanced by these high-velocity microjets, in turn preventing fouling of the electrode surface and enhancing electron transport processes [14].

In water electrolysis, the cell voltage ($V_{\text{cell}}$) is a crucial factor that represents energy consumption. The practical cell voltage is expressed in Eq. (7.1), where $E_a$ is the anode potential for the $O_2$ evolution reaction (OER), $E_c$ is cathode potential for the $H_2$ evolution reaction (HER), $j$ is the current density, $\sum R$ is the total ohmic resistance, $E^{\text{rev}}$ is the reversible potential, $\eta_a$ is the anode overpotential and $\eta_c$ is the cathode overpotential [15].

$$V_{\text{cell}} = |E_c| - |E_a| + j \times \sum R = E^{\text{rev}} + |\eta_a| - |\eta_c| + j \times \sum R \tag{7.1}$$

Based on Eq. (7.1), the total cell voltage ($V_{\text{cell}}$) is greatly influenced by the anodic and cathodic overpotentials, and the ohmic voltage. Therefore, industrial $H_2$ production by water electrolysis aims at reducing the ohmic voltage and overall overpotentials. It has been recently demonstrated experimentally that power ultrasound significantly reduces the reaction overpotential and ohmic voltage [16].

## Effect of Ultrasound on the Hydrogen Evolution Reaction and the Oxygen Evolution Reaction

Reactions occurring at gas-evolving electrodes constitute an essential area of electrochemical engineering, especially in water electrolysis for $H_2$ production, electrowinning of metals and corrosion. In the case of water electrolysis, the cell voltage ($V_{cell}$) includes a term $j \times \sum R$ due to the ohmic drop across the cell (see Eq. 7.1), which is increased by gas bubbles adhering to the electrode surface. Usually, the ohmic contribution is minimised with a suitable geometric configuration of the electrodes and the use of highly conducting electrolytes. Despite this, $H_2$ $O_2$ gases possess insulating properties, so their presence on the electrode surface reduces the electrodes effective area and contributes to an increase in $j \times \sum R$. There are also effects of the attached gas bubbles onto the surface and concentration overpotentials, which were discussed by Dukovic and Tobias [17].

Walton et al. [18] studied the effect of ultrasound on $H_2$ evolution from a 1 M $H_2SO_4$ solution at a platinised platinum electrode under 38 kHz insonation. They observed an increase in current of 2.1-fold with the presence of ultrasound compared to silent (no ultrasound) conditions. They found that, although the reduction of the proton ($H^+$) at the platinised platinum is a "reversible" reaction, the availability of $H^+$ at the electrode did not improve the current caused by enhanced diffusion. Instead, they concluded that the dominant effect in $H_2$ evolution was the removal of $H_2$ bubbles from the electrode surface [18].

McMurray et al. [19] showed that, by careful electrochemical experiments and analyses (Tafel), the rates of the electrochemical $O_2$ reduction and $H_2$ evolution at a vibrating titanium electrode are significantly increased with power ultrasound (20 kHz, 26 W/cm$^{-2}$). They attributed their findings for the OER and HER to the enhanced mass transport and activation-controlled processes, respectively.

Ultrasound can also affect the electrode overpotential of a process and reports by Moriguchi [20] and Pollet et al. [21] showed that the overpotential of OER and HER on Ag (silver), Pt and SS (stainless-steel) electrodes immersed in aqueous solutions decreases upon sonication. In the same paper, Pollet et al. [21] showed that under-sonication at 20 and 500 kHz, the reduction wave shifted anodically with increasing ultrasonic intensity. Similarly, the discharge of $H_2$ and $O_2$ shifted anodically and cathodically, respectively, with increasing ultrasonic power. This decrease in decomposition voltages in the presence of ultrasound was found to be mainly due to the combined effect of a decrease in anodic and cathodic overpotentials (absolute values). Moreover, it was found that the decrease in overpotential occurred without any appreciable change in the Tafel slopes suggesting that the electron-transfer coefficient ($\alpha$), was not affected by sonication. In other words, the partitioning of the change in the potential energy of the system between the forward and reverse reactions were not affected by insonation; however, by carefully examining the Tafel plots, the exchange current density ($i_o$) values obtained were different in the absence and presence of ultrasound. The Tafel equation is given by:

$$\eta = a + b\ \log i_{\text{net}} \tag{7.2}$$

where $a = (2.3\ \text{RT}\ \log i_0)/(\alpha nF)$; and, $b = -2.3\ \text{RT}/(\alpha nF)$. The decrease in the $a$ values (i.e. intercepts) was observed to be influenced by ultrasonic intensity. Since the electron-transfer coefficient appeared not to be affected by sonication, this change in $a$ was due mainly to an increase in exchange current density. In their conditions, the exchange current density at maximum ultrasonic intensity was found to be 300% higher than under silent conditions.

Since the exchange current density is proportional to the apparent heterogeneous rate constant ($k^o$), this finding suggested that the apparent heterogeneous rate constant was also increased by the same amount in the presence of ultrasound. It was suggested by Pollet et al. [21] that this increase in the apparent heterogeneous rate constant was due to either changes in electrode surface composition or changes in temperature. Since the macroscopic temperature of the bulk did not change during sonication, the increase in the apparent heterogeneous rate constant was possibly due to either a modification of the electrode surface composition or an increase in the microscopic temperature within the diffusion layer due to the implosion of high-energy cavitation bubbles.

Pollet et al. [21] proposed two mechanisms to explain the decrease in overpotential under conditions where concentration polarisation involving the electrolyte was negligible. The first was that ultrasonically produced cavitation modifies the surface of the electrode, for example, by changing the number of sites available for the adsorption of $H_2$ ($H_{ad}$) on the electrode surface. The erosive action associated with the implosion of high-energy cavitation bubbles produced a new electrode surface continuously and at the same time promoted the removal of adsorbed impurities from the electrode surface. Also, it is known that the apparent heterogeneous rate constant depends upon the overpotential, which in turn depends upon the active sites available on the electrode surface for the electron transfer. Since a decrease in overpotential led to an increase in active sites, it was speculated that, on the application of ultrasound, the electron transfer became more superficial.

The second mechanism proposed to explain a part of the decrease in $H_2$ overpotential ($\eta_{\text{Hydrogen}}$) produced by ultrasound involves the degassing effects associated with microstreaming together with cavitation. As highlighted in other articles, it is well known that the solution adjacent to the electrode surface is supersaturated with molecular $H_2$ (because of the low solubility of molecular $H_2$ in aqueous solutions) leading to the so-called "bubble overpotential." It is also known that acoustic streaming and cavitation help degas the solution immediately adjacent to the electrode, therefore decreasing and even eliminating the "bubble overpotential." In other words, they found that both $H_2$ bubble ($\eta_{\text{Hydrogen}}$) and $O_2$ bubble ($\eta_{\text{Oxygen}}$) overpotentials significantly decreased under sonication. The effect of ultrasound on cell potential, efficiency and a specific energy for $H_2$ production by water electrolysis is illustrated in Fig. 7.3.

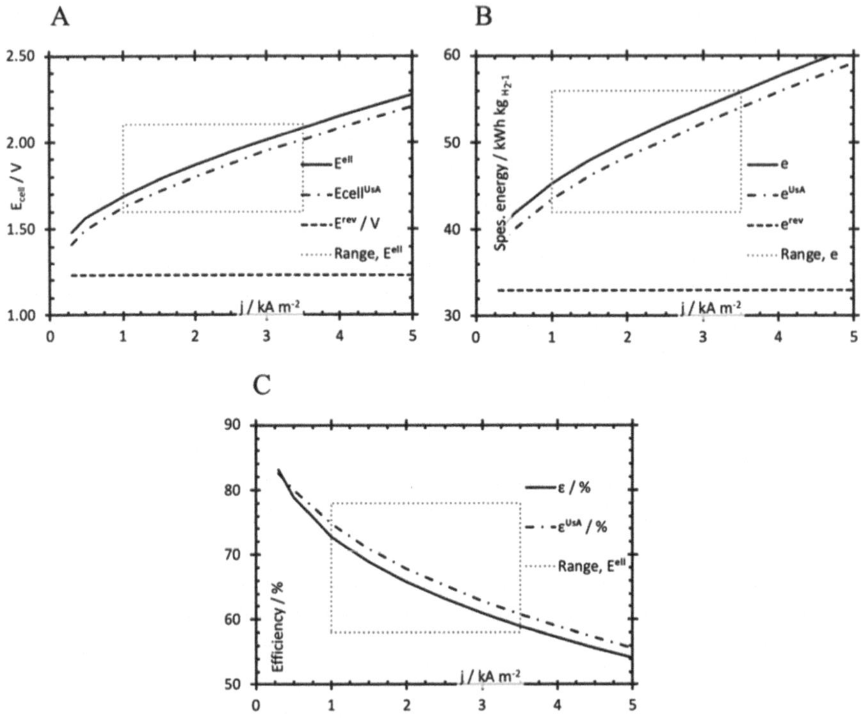


**Fig. 7.3** Effect of ultrasound on (**a**) cell potential, (**b**) efficiency and (**c**) specific energy for $H_2$ production (UsA = Ultrasound-Assisted)

## Effect of Ultrasound upon Hydrogen Yield

Ultrasound-assisted water electrolysis for $H_2$ production was first carried out by Cataldo [22], who studied the effect of ultrasound (30 kHz and 1–2 W cm$^{-2}$) upon the yield of gases from a saturated aqueous solution of NaCl (6.0 M), HCl (6.0 M) and acidified NaCl (5.0 M NaCl/1.1 M HCl) using both platinum and carbon rods as electrodes. He found that ultrasound dramatically increases the yields of chlorine ($Cl_2$) and $H_2$ gases. The strong degassing effect at the electrode surface due to "bubble fusion" caused by cavitation and efficient stirring is the most crucial reason for the enhanced gas yield [22].

Lepesant and other researchers such as Zadeh [16] and Symes [23], under the supervision of Pollet, investigated the effect of ultrasound upon the electrolytic $H_2$ production from weak acidic ($H_2SO_4$) and alkaline (NaOH and KOH) solutions using various electrode materials (platinum—Pt, glassy carbon—GC, industrial carbon—C and 316 stainless steel—316SS). Overall, they found that ultrasound dramatically reduces the overall reaction overpotential by 100–400 mV depending upon the solution type, electrolyte concentration, electrode material and ultrasonic intensity used. The thermodynamic decomposition voltage of water is +1.229 V vs. RHE [24], as shown:

$$W = \frac{nF}{3600 \times Vm \times 1000} \times V_{\text{cell}} = \frac{2 \times 96485}{3600 \times 22.675} \times V_{\text{cell}}$$
$$= 2.3639 \times V_{\text{cell}}(\text{kWh}) \quad (7.3)$$

The theoretical energy consumption for producing 1 $m^3$ of $H_2$ is 2.90 kWh/$m^3$ of $H_2$ [15]; however, practically, a water electrolytic cell usually operates between +1.70 and +2.5 V. In other words, the practical energy consumption is much higher than the theoretical energy consumption.

If one assumes a working cell voltage of a water electrolysis cell close to +2.0 V, then this leads to energy consumption for $H_2$ production higher than 50 kWh/kg (i.e. higher than the theoretical value under standard conditions of 33 kWh/kg) [1].

In the case of water electrolysis of 0.1 M KOH under *STP* at Pt electrodes, the practical energy consumption to produce 1 $m^3$ of $H_2$ is 5.96 kWh (@ +1.70 V). It was found that by using ultrasound, the practical energy consumption may be reduced to 5.0587 kWh. Therefore, a decrease of around 14% of energy consumption and 10% of energy efficiency can be achieved by using sonication in alkaline water electrolysis [16]. A summary of the energy efficiency calculations for producing 1 $m^3$ of $H_2$ by electrolysis of 0.1 M KOH are explained below.

Theoretical energy consumption based on Eq. (7.3) is:

$$W_t = 2.3639 \times 1.229 = 2.90 \text{ kWh} \quad (7.4)$$

Practical energy consumption when the decomposition voltage is equal to 2.52 V [16] is:

$$W_{\text{p}} = 2.3639 \times 2.52 = 5.96 \text{ kWh} \quad (7.5)$$

By using ultrasound, the decomposition voltage reduces to 2.14 V [16]. Therefore, electrochemical energy consumption by the aid of ultrasound is:

$$W_{\text{UsA}} = 2.3639 \times 2.14 = 5.0587 \text{ kWh} \quad (7.6)$$

The amount of energy ($W_{\text{Us}}$) consumed by an ultrasonic transducer of 200 W working with 30% amplitude is:

$$W_{\text{Us}} = 200 \text{ W} \times 0.3 + 25.3 \text{ W} = 85.3 \text{ W} = 0.0853 \text{ kWh} \quad (7.7)$$

where 25.3 W is the idle energy consumed by the ultrasonic generator. Therefore, the electrochemical energy efficiency is:

$$\eta_{\text{Elect}} = \frac{W_{\text{t}}}{W_{\text{p}}} \times 100 = \frac{2.90}{5.96} \times 100 = 48.74\% \quad (7.8)$$

Moreover, the ultrasound-aided electrochemical energy efficiency is:

$$\eta_{\mathrm{UsA+Elect}} = \frac{W_{\mathrm{t}}}{W_{\mathrm{UsA}} W_{\mathrm{Us}}} = \frac{2.90}{5.0587 + 0.0853} = 56.37\% \tag{7.9}$$

The type of solutions and the concentration have a significant effect on the $H_2$ production by electrolysis. For example, the yield of $H_2$ is significantly higher for acidified sodium chloride (NaCl) solutions than saturated ones. According to Zadeh [16], the electrolytic $H_2$ production was improved by 14 and 25% for NaOH and KOH solutions, respectively, during sonication at similar concentrations. The higher production rate of $H_2$ from KOH than NaOH was caused by the effect of ultrasound on solution conductivity [16].

Li et al. [25] studied the effect of ultrasound (60 kHz and 50 W) for water electrolysis in various electrolyte concentrations of 0.1 M, 0.5 M and 1.0 M NaOH solutions. It was observed that the cell voltage, anode and cathode potential were significantly decreased at higher electrolyte concentrations. This was due to an increase in electrolyte conductivity. The electrolyte concentration influences the $H_2$ bubble size and applied cell voltage. Generally, the bubble size becomes smaller with increasing electrolyte concentration. Li et al. [25] observed that the efficiency of $H_2$ generation in the absence of ultrasound was found to be in the range of 60–75%. On the other hand, in the presence of ultrasound, the efficiency was improved significantly and found to be in the range of 80–85% [25].

## Summary and Outlook

Worldwide, many research activities are being carried out towards the improvement of existing and the development of novel $H_2$ production technologies. Currently, the most widely used and technically well-proven $H_2$ method is the reforming of hydrocarbons; however, this method heavily relies upon the use of fossil fuels, in turn yielding significant greenhouse gas emissions.

There are other alternative routes for producing greener $H_2$. For example, $H_2$ produced from acidic and alkaline water electrolysis (powered by renewable technologies) could be a solution. Another method could be the use of ultrasound in the presence of an electrolytic cell voltage (and current). Sonoelectrochemical $H_2$ production methods have not been fully explored, and some critical parameters such as the influence of the ultrasonic frequency, ultrasonic power, electrolyte type and concentration and electrode materials, need to be investigated. Another area that requires more considerable attention is the quantitation of the produced $H_2$. Detailed quantitation is necessary in order to understand the effect of varying operating conditions as well as upgrading the process from laboratory to pilot or industrial scale (as long as a techno-economic analysis for an industrial application is done). Moreover, to the best of our knowledge, no reports are investigating the sonoelectrochemical production of $H_2$ from non-aqueous solutions.

**Acknowledgements** The authors are grateful to the ENERSENSE programme and NTNU Team Hydrogen at the Norwegian University of Science and Technology (NTNU) for supporting and helping on this book project.

## References

1. Coutanceau C, Baranton S, Audichon T (2018) Hydrogen electrochemical production. In: Pollet BG (ed) Hydrogen electrochemical production. Academic, Cambridge, 1–6 p
2. Pollet BG (2012) Power ultrasound in electrochemistry: from versatile laboratory tool to engineering solution. Wiley, Chichester, 1–344 p
3. Lamb JJ, Islam MH, Hjelme DR, Pollet BG, Lien KM (2019) Effect of power ultrasound and Fenton reagents on the biomethane potential from steam-exploded birchwood. Ultrasonics Sonochem 58:104675
4. Islam MH, Lamb JJ, Lien KM, Burheim OS, Hihn J-Y, Pollet BG (2019) Novel fuel production based on sonochemistry and sonoelectrochemistry. Meeting Abstracts - The Electrochemical Society [ECS] ( ISSN 1091-8213)
5. Merouani S, Hamdaoui O (2016) The size of active bubbles for the production of hydrogen in sonochemical reaction field. Ultrason Sonochem 32:320–327
6. Pollet BG (2012) Power ultrasound in electrochemistry: from versatile laboratory tool to engineering solution. In: Pollet BG (ed) Power ultrasound in electrochemistry: from versatile laboratory tool to engineering solution. Wiley, Chichester, 1–344 p
7. Kalamaras CM, Efstathiou AM (2013) Hydrogen production technologies: current state and future developments. Conf Pap Energy 2013:9
8. Demirbas MF (2006) Hydrogen from various biomass species via pyrolysis and steam gasification processes. Energy Sources, Part A Recover Util Environ Eff 28(3):245–252
9. Levin DB, Pitt L, Love M (2004) Biohydrogen production: prospects and limitations to practical application. Int J Hydrog Energy 29(2):173–185
10. Lamb JJ, Pollet BG, Burheim OS (2020) Energy storage. In: Energy-smart buildings design: construction and monitoring of buildings for improved energy efficiency. https://doi.org/10.1088/978-0-7503-3259-0ch6
11. Ivy J (2004) Summary of electrolytic hydrogen production milestone completion report. National Renewable Energy Laboratory
12. Hujjatul Islam M, Paul MTY, Burheim OS, Pollet BG (2019) Recent developments in the sonoelectrochemical synthesis of nanomaterials. Ultrason Sonochem 59:104711
13. Pollet BG, Hihn J-Y, Doche M-L, Lorimer JP, Mandroyan A, Mason TJ (2007) Transport limited currents close to an ultrasonic horn. J Electrochem Soc 154(10):E131
14. Mason TJ (1990) Sonochemistry: the uses of ultrasound in chemistry. Royal Society of Chemistry, Cambridge
15. Wang M, Wang Z, Gong X, Guo Z (2014) The intensification technologies to water electrolysis for hydrogen production – a review. Renew Sust Energ Rev 29:573–588
16. Zadeh SH (2014) Hydrogen production via ultrasound-aided alkaline water electrolysis. J Autom Control Eng 2(1):103–109
17. Dukovic J, Tobias CW (1987) The influence of attached bubbles on potential drop and current distribution at gas-evolving electrodes. J Electrochem Soc 134(2):331
18. Walton DJ, Burke LD, Murphy MM (1996) Sonoelectrochemistry: and oxygen evolution chlorine, hydrogen at platinised platinum. Electrochim Acta 41(17):2747–2751
19. McMurray HN (1998) Hydrogen evolution and oxygen reduction at a titanium sonotrode. Chem Commun 8:887–888
20. Moriguchi N (1934) The influence of supersonic waves on chemical phenomena. III The influence on the concentration polarisation. Nippon Kagaku Kaishi 55(8):749–750

21. Pollet B, Lorimer JP, Phull SS, Mason TJ, Walton DJ, Hihn JY et al (1999) The effect of ultrasonic frequency and intensity upon electrode kinetic parameters for the Ag(S2O3)23−/Ag redox couple. J Appl Electrochem 29(12):1359–1366
22. Cataldo F (1992) Effects of ultrasound on the yield of hydrogen and chlorine during electrolysis of aqueous solutions of NaCl or HCl. J Electroanal Chem 332(1–2):325–331
23. Symes D (2011) Sonoelectrochemical (20 kHz) production of hydrogen from aqueous solutions. The University of Birmingham, Birmingham
24. González-Cobos J, Baranton S, Coutanceau C (2016) Development of bi-modified PtPd nanocatalysts for the electrochemical reforming of polyols into hydrogen and value-added chemicals. ChemElectroChem 3:1694–1704
25. De Li S, Wang CC, Chen CY (2009) Water electrolysis in the presence of an ultrasonic field. Electrochim Acta 54(15):3877–3883

# Low-Grade Waste Heat to Hydrogen

# 8

Yash D. Raka, Robert Bock, Jacob J. Lamb, Bruno G. Pollet, and Odne S. Burheim

## Introduction

Industrial waste heat is a source of energy that is currently not fully utilised. On a global scale, the total amount of waste heat accounts for more than 40% of the total energy use, meaning almost half of the energy consumed is wasted as heat [1]. The waste heat potential in the EU has been estimated to be 304 TWh/year [2]. Utilising this waste heat provides economic and environmental benefits. For example, in 2013, it was estimated to cost about 370 million USD for reducing $CO_2$ equivalent ($CO_2$ eq) emissions from waste heat from industries in the United Kingdom [3]. The EU addressed this issue with a policy that recommends a reduction in GHG emissions by 40% and improvements to energy efficiency by 27% in the transportation and industrial sectors by 2030. Also, the use of $H_2$ in the transportation sector was identified to be an alternative solution that caters to both energy efficiency and reduced $CO_2$ emissions.

Y. D. Raka · R. Bock
Department of Energy and Process Engineering, ENERSENSE, Norwegian University of Science and Technology, Trondheim, Norway
e-mail: yash.raka@ntnu.no; robert.bock@ntnu.no

J. J. Lamb (✉)
Department of Electronic Systems, ENERSENSE, Norwegian University of Science and Technology, Trondheim, Norway

Department of Energy and Process Engineering, ENERSENSE, Norwegian University of Science and Technology, Trondheim, Norway
e-mail: jacob.j.lamb@ntnu.no

B. G. Pollet · O. S. Burheim
Department of Energy and Process Engineering, ENERSENSE and NTNU Team Hydrogen, Norwegian University of Science and Technology, Trondheim, Norway
e-mail: bruno.g.pollet@ntnu.no; odne.s.burheim@ntnu.no



J. J. Lamb, B. G. Pollet (eds.), *Micro-Optics and Energy*,
https://doi.org/10.1007/978-3-030-43676-6_8

The quality of waste heat depends on its temperature. Heat available at temperatures below 373 K is often referred to as low-grade waste heat. A considerable amount of low-grade waste heat is available worldwide in the temperature range between 323 and 373 K (e.g. 833 TWh per annum in the United States alone) [4]. This heat cannot be reintegrated entirely on-site or be used for district heating. The Organic Rankine Cycle (ORC) and thermoelectric generators are some of the few technologies that can utilise this energy. ORC has an intrinsic limitation of efficiency like that of the Carnot cycle, where the efficient operating temperature range for the power plants using this cycle is above 373 K [5].

Furthermore, energy recovery using ORC has added disadvantages, such as environmentally harmful refrigerants in some cases and limitations for downsizing. While thermoelectric generators have no moving parts, their low efficiency and poor form factor make them a less appropriate choice [5]. To date, none of these technologies have demonstrated conversion of heat to $H_2$ at temperatures below 373 K with efficiencies and costs adhering to industrial standards.

Reverse electrodialysis (RED) is a membrane-based technology that can convert a chemical potential into electricity directly by controlling the mixing of two different salinity solutions with ion-exchange membranes (IEM). A RED stack consists of two electrodes together with alternately stacked anion exchange membranes (AEM) and cation exchange membranes (CEM) separated by spacers. When high-concentration (HC) and low-concentration (LC) solutions are fed into the stack, cations in the HC will migrate through the CEMs and anions migrate through the AEMs into LC due to the concentration gradient. This ionic flux is converted into an electron flux at the electrodes via redox reactions, generating an electric current for supplying an external load. RED with freshwater and seawater as feed solution does not emit environmentally toxic gas (chlorine gas evolution may take place) and has inexhaustible access to free fuel supply in the vicinity of sites where freshwater and seawater meet; however, accessibility to these natural resources limits its locational options.

One of the barriers for commercialising RED systems is the high maintenance cost of the IEMs, which need to be either replaced with new ones or recovered when affected by different types of fouling. These disadvantages can be overcome by using an artificial saline solution that ensures the availability of large amounts of Gibbs free energy of mixing, low energy consumption in the regeneration unit to restore the original concentration levels of feed solutions and use of waste heat at low temperature. The different salinity solutions provide the driving force by separating the increased solute from the diluted side and reintroducing it in the concentrated side without any fresh solution supply. The closed-loop system mitigates the geographical limitations since there is no need for a continuous supply of freshwater and seawater, preventing the danger of membrane fouling as well as saving pretreatment costs [6].

## Theoretical Background

### Regeneration Process

In the RED stack, salt is transferred from the concentrated to the dilute solution. At the same time, the solvent is transferred from the dilute to the concentrated solution due to osmotic pressure across the ionic exchange membrane. This results in a different salt concentration of the solutions exiting the RED stack than those entering the stack. The regeneration stage restores the exiting solutions to the concentrations of the feed solutions. This can be done by extracting either the solvent or the salt from the solutions that leave the RED system. Either of the components extracted is recycled to restore the solutions to meet the desired concentrations. There are two different ways to restore the concentrations, as outlined in the process flow diagrams in Fig. 8.1.

In the solvent extraction process, the amount of salt transferred in the RED stack from the concentrated to the dilute solution is rebalanced by the addition of stream 1 to 2 followed by the necessary solvent excess extraction. The two exiting streams are the solute-rich stream (stream 5) and a solute-free stream (stream 4). Stream

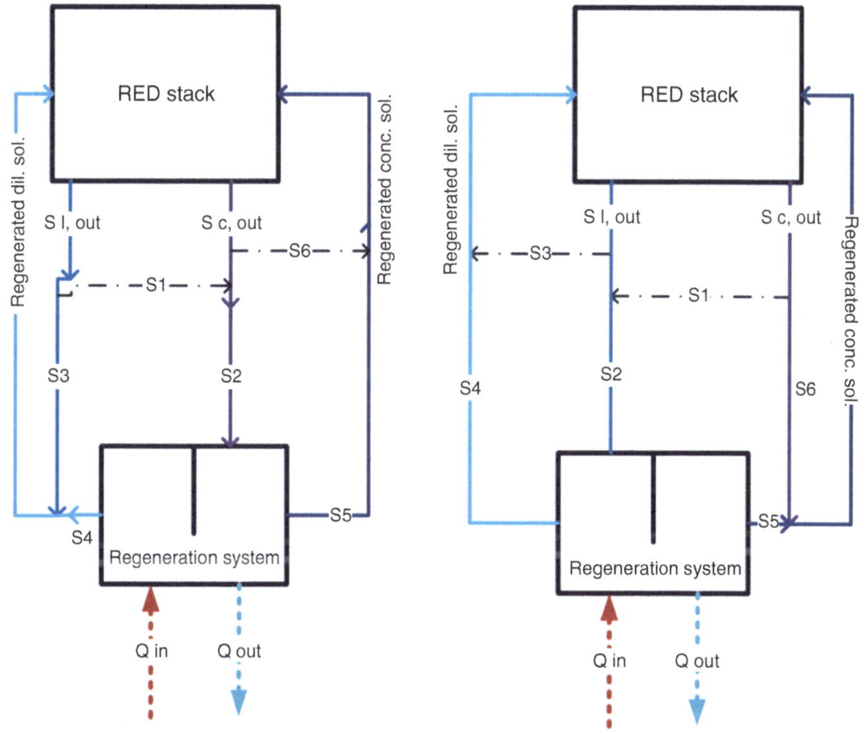


**Fig. 8.1** Schematic of reverse electrodialysis (RED) with solvent extraction regeneration process (left) and solute extraction regeneration process (right)

6 reduces the input flow rate to the regeneration stage, resulting in a decrease of heating requirement.

In the salt extraction process, the amount of salt transferred in the RED stack from the concentrate solution to the diluted one is extracted from the dilute stream 2 in the regeneration unit. The output streams 4 and 5 are solvent rich and solute abundant streams. Thermolytic salts, for example, ammonium bicarbonate, decompose into the gaseous compounds $NH_3$ and $CO_2$ by increasing the temperature of the solution at and above 333 K. These gases are subsequently reabsorbed in the concentrate solution, thus restoring it to the initial concentration as shown in Fig. 8.2. Such thermolytic solutions have the potential to utilise a low-grade heat source in the temperature range of 323–333 K. Ammonium bicarbonate ($NH_4HCO_3$) salt is one of the promising thermolytic salt solutions. The low decomposition temperature of around 333 K at 1 atm, high solubility in water and relatively low molecular weight make it a suitable choice for the thermolytic salt needed in this process.

Luo et al. performed initial RED experiments using $NH_4HCO_3$ with varying operating conditions such as flow rate and feed solution concentration. They showed that the internal resistance and open circuit potential decrease with an increase in the dilute solution concentration. Further, an increase in flow rate increases the open circuit potential while decreasing the internal resistance as well [7]. Hatzell et al. demonstrated the influence of spacers and channel alignment on the internal resistance and the power density. They found that reorienting the flow through the channels decreased the internal resistance due to decreased bubble coverage by 13%; reducing channel length increased the limiting current density and the power density by 23% and 14%, respectively, and spacer-less stacks had lower ohmic resistance than those with spacers but at low flow rates [8]. Zhu et al. compared NaCl and $NH_4HCO_3$ salts and found that at the same concentration ratios the energy recovery and power density are lower for $NH_4HCO_3$ than for NaCl mainly due to the low conductivity (low ion activity) [9]. Kwon et al. studied parameters such as the concentration difference, the membrane type, the inlet flow rate and the compartment thickness. Their results indicate that the open circuit potential highly depends on the concentration difference across an IEM, and the maximum power density increases with increasing inlet flow rate and decreasing the intermembrane distance [10].

Bevacqua et al. performed a parametric study evaluating parameters such as dilute solution concentration and feed velocity. They observed that for feed velocities higher than 1 cm/s, the gross power density reduces to a negative power output owing to losses due to a more significant pressure drop [11].

Hatzell et al. did the first study for $H_2$ production using RED. They studied different cathode reactions [$H_2$ evolution reaction (HER) and $O_2$ reduction reaction (ORR)] and catalysts for energy production from $NH_4HCO_3$ RED. They found that the potential for direct $H_2$ production at limiting current density using an HER cathode gives the highest energy, 1.5 times (most in the form of $H_2$) the energy produced by cathodes with ORR. In contrast, the ORR cathode produces more electricity. Further, indirect $H_2$ production (i.e. coupling RED and electrolysis) resulted in one-third energy to that of direct $H_2$ production [12]. Nazemi et al.

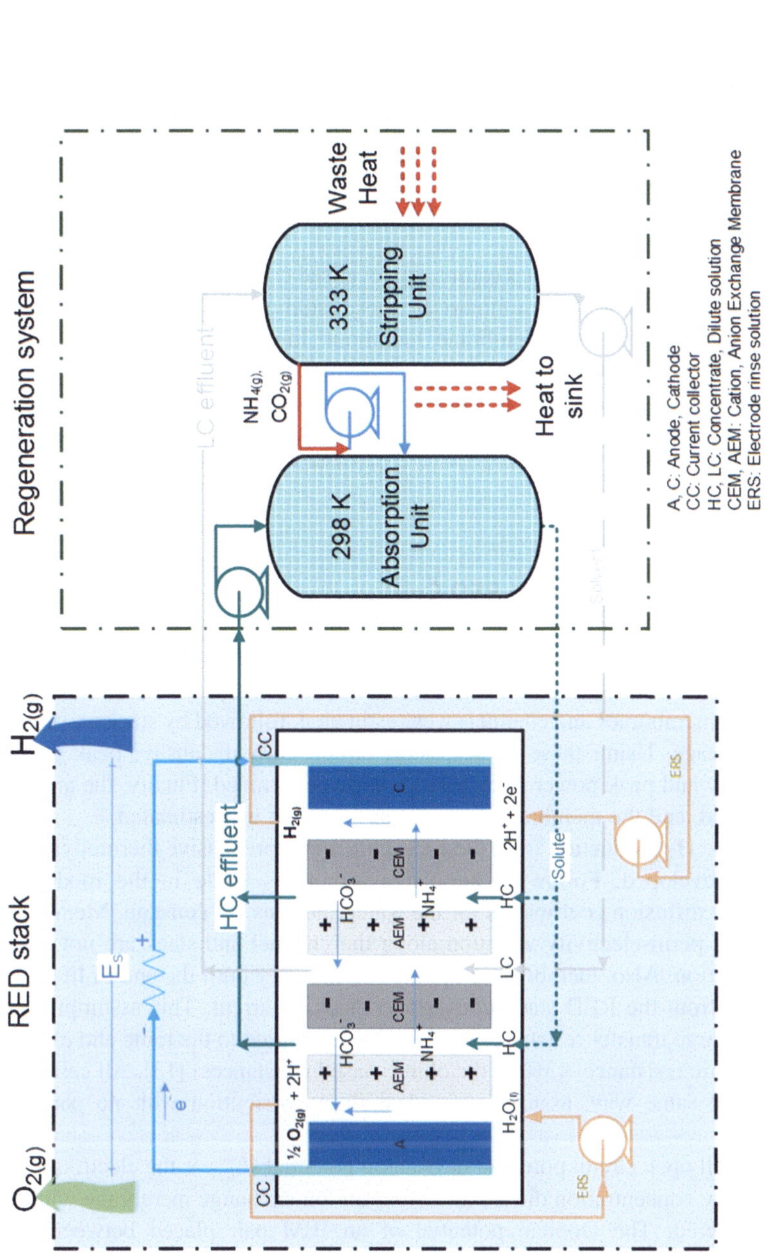


**Fig. 8.2** Schematic of reverse electrodialysis (RED) system for ammonium bicarbonate salt with solute extraction regeneration system for $H_2$ production

developed a thermodynamic model to evaluate how much electricity and $H_2$ gas can be extracted from the natural mixing of river and seawater. The results indicate that the maximum current density and $H_2$ production is achieved with five cell pairs, while the maximum electrical power density was reached with 20 cell pairs [13]. Chen et al. demonstrated the use of acidic catholyte and alkaline anolyte for $H_2$ and $O_2$ production in a RED stack. They found that the current density (concentration difference) and the electrode rinse solution's concentration ($H_2$ evolution over-potential) has a strong influence on $H_2$ production [14]. All these studies have been summarised in Table 8.1.

There have been several studies estimating levelised cost of electricity (LCoE) for RED technology. Table 8.2 summarises these studies indicating diversity in parameters such as cost of the membrane, membrane life, power density and LCoE.

In the present study, we develop a detailed thermodynamic model to evaluate $H_2$ production by using reverse electrodialysis of ammonium bicarbonate. Further, a parametric study showing the influence of various parameters such as the feed solution concentrations, the electrochemical properties of the membrane and the cell geometry is presented.

## Thermodynamic Model of a RED Cell

At first, the input data, as shown in Table 8.3, were initialised. Parameters such as conductivity and activity coefficients were estimated. On this basis, cell voltage, resistances of membranes and channels were estimated, followed by stack resistance and stack voltage. Using these values, short circuit current density, peak power current density and peak power stack voltage were determined. Finally, the amount of $H_2$ produced, and the membrane area required for that was estimated.

To calculate $H_2$ production from the RED cell, a comprehensive thermodynamic model was developed. Following are the assumptions made in the modelling approach. The diffusion coefficients for the water and ions are constant. Membrane resistance and permselectivity variation along the channel and stack are not taken into consideration. Also, membrane properties do not vary until the end of life. The ionic current from the RED stack converts to electric current. This assumption is valid if the charge transfer resistance is negligible compared to the ionic and electric resistance (Ionic resistance $<3\%$ of the ohmic-based resistances) [13]. All cell pairs operate in the same way, assuming an ideal flow distribution with no parasitic currents.

The unit cell open circuit potential or Donnan potential ($E_{\mathrm{u.c}}^{\mathrm{oc}}$) is the electromotive force driven by concentration difference across an ion exchange membrane with no losses considered. The Donnan potential of an IEM pair placed between two different concentration solutions can be described using the modified Nernst equation:

**Table 8.1** Ammonium bicarbonate reverse electrodialysis systems (AmB RED) and hydrogen production from RED

| | Concentration (M) | | Membrane | | | | | | | Flow rates | | | | | |
|---|---|---|---|---|---|---|---|---|---|---|---|---|---|---|---|
| Type | Concentrated | Dilute | Name | Permselectivity | Ohmic resistance (Ω/m$^2$) | Area (cm$^2$) | Thickness (μm) | Channel thickness (μm) | No. of cell pairs | Feed (ml/min) | Redox (ml/min) | Electrolyte-electrode | Potential | Power | References |
| Experimental | 1.5 | 0.003–0.15 | Selemion CMV and AMV | 1 | na | 10.5 × 7.5 | 130 | 500 | 20 | 150–1200 | 5 | 0.1 M $K_4Fe(CN)_6$ and $K_3Fe(CN)_6$ (50:50) Titanium plates coated with Ru–Ir | 3.7 | 0.33 | [7] |
| | 0.6–1.5 | 0.005–0.02 | Neosepta CMX and AMX | na | na | 7 × 7 | na | 200–800 | | 1–20 | 50 | 0.05 M $K_4Fe(CN)_6$ and $K_3Fe(CN)_6$ Ru–Ir coated titanium electrodes | 1 | 0.62 | [10] |
| | | | Excellion I-100 and I-200 | | | | | | | | | 0.3 M $NH_4HCO_3$ Ru–Ir coated titanium electrodes | | | |
| | 2 | 0.01–0.1 | Fujifilm RP1 80050-04, 80045-01 | na | na | 10 × 10 | 120 | 270 | 10 | 0.5–2 cm/s | 225 | 0.1 M $K_4Fe(CN)_6$ and $K_3Fe(CN)_6 \cdot 3H_2O$ Ru–Ir oxide-coated Ti electrode | 1.56 | 2.42 | [11] |
| | | | Nafion at electrode | | | | | | | | | 1 M $NH_4HCO_3$ Ru–Ir oxide-coated Ti electrodes | | | |
| | 1.1 | 0.011 | Selemion CMV and AMV | Na | Na | 2 × 1 | na | 1100 | 1 | 1–18 mm/s | | Pt mesh | 0.155 | 0.15 | [8] |
| | 1.5 and 0.6 | 0.015 and 0.006 | PC-SK and SA | Na | Na | 8 × 8 | na | 500 | 10 | 100 | | Ti mesh coated with Pt–Ir | | 0.32 | [9] |

(continued)

**Table 8.1** (continued)

| | Concentration (M) | | Membrane | | | | | | | Flow rates | | | | | |
|---|---|---|---|---|---|---|---|---|---|---|---|---|---|---|---|
| Type | Concentrated | Dilute | Name | Permselectivity | Ohmic resistance ($\Omega/m^2$) | Area ($cm^2$) | Thickness (μm) | Channel thickness (μm) | No. of cell pairs | Feed (ml/min) | Redox (ml/min) | Electrolyte-electrode | Potential | Power | References |
| Theoretical | 0.05–2 | 0.01 | CMV/ AMV | 0.88–0.95 | | 49 | | 100–1000 | | 0.1–20 | | 0.05 M $K_4Fe(CN)_6$ and $K_3Fe(CN)_6$<br>Ti coated with Ir and Ru | 1.1 | 0.84 | [15] |
| | | | | 0.86–0.96 | | | | | | | | Ti coated with Ir and Ru | | | |
| | 0.5–2.6 | 0.01–0.1 | Fujifilm | 0.754 | 2E-4 $(Clow)^{-}_{0.236}$ | 50 × 50 | | 270 | 1000 | 0.5–5 cm/s | | | 1.8 | 7.5 | [16] |
| | | | | | 4E-5 $(Clow)^{-}_{0.236}$ | | | | | | | | | | |
| | 1–2.5 | 0.01–2 | | 0.9 | 0.5 | 10 × 10 | | 270 | 10 | 0.1–3 cm/s | | | | | |
| | 0.5–2 | 0.01–0.1 | | 0.85 and 0.95 | $6 \times 10^{-4}$ to $1.5 \times 10^{-4}$ | 50 × 50 | | | 1000 | 0.5 to 2 | | | | | |
| | 0.6 | 0.0015 | | 1 | 3 | 0.635 | | 150 | 5 to 20 | | | | | 1.72 Mol/ $cm^2$/h | [13] |
| Hydrogen | 1.5 | | | | | 207 | 250 | 20 | 10 to 20 | 400 | 100 | Ti mesh coated with Pt/Ir<br>Pt/C on carbon cloth-ORR and HER<br>Pt/Ir on Ti or MoB/C on carbon cloth for only HER | | 118 Wh/ $m^3$ | [12] |
| | 1.5–5 | 0.008–0.5 | CJMC-1 CJMA-1 | | | 9 × 21 | 160 to 200 | 750 | | Na | 75 | 0.2–2 M HCl and 0.5 NaOH and 0.5 $Na_2SO_4$ | 2.91 | 0.72 ml/ $cm^2$/h | [14] |

**Table 8.2** Summary of previous economic studies on RED for electricity production

| Author | Year | Capacity (MW) | Plant life (year) | $C_{mem}$ (€/m$^2$) | $t_{mem}$ (year) | $P$ (W/m$^2$) | LCOE (€/kWh) | Type |
|---|---|---|---|---|---|---|---|---|
| Turek [17] | 2007 | – | – | 100[a,b] | 10 | 0.46 | 6.79[a] | River/sea |
| Post [18] | 2010 | 200 | 40 | 2 | 7 | 2 | 0.08 | River/sea |
| Daniilidis [19] | 2014 | 200 | 25 | 50 | 7 | 2.2 | 0.72 | River/sea |
| Weiner [20] | 2014 | – | 20 | 750[a,b] | – | 1.2 | 6.33[a] | River/sea |
| Bevacqua [16] | 2017 | 100 | 20 | 50 | 20 | 4.78 | 0.3 | AmB |
| Micari [21] | 2019 | 1 | 30 | 1–30 | – | | 0.4 | RED-HE |
| Giacalone [22] | 2019 | 0.001–1000 | – | 4, 15 | 10 | >1–19 | 0.27–0.33 | River/sea |
| Papapetrou [23] | 2019 | 0.5–50 | 30 | 1–30 | 10 | 0.8–33 | 1.36–3.69 | RED-HE |
| Raka [24] | 2019 | 1500 $kg_{H2}$ | 20 | 3, 4, 150 | 4, 7, 10 | 0.7–3.7 | 1.71–153 $kg_{H2}$ | RED-HE-H2 |

[a]Currency in $

[b]Cost includes membranes, gasket, spacers, electrodes and end plate

**Table 8.3** Input thermodynamic and economic data for reverse electrodialysis

| Parameter | Symbols | Value | Unit |
|---|---|---|---|
| Concentrate solution conc. | $c_{\text{hc}}$ | 2.6 | M |
| Dilute solution conc. | $c_{\text{lc}}$ | 0.1 | M |
| Ideal gas constant | $R$ | 8.314 | J/K/mol |
| Temperature | $T$ | 293 | K |
| Faraday's constant | $F$ | 96,485 | C/mol |
| Permselectivity of CEM/AEM | $\alpha$ | 0.8 | |
| Diffusivity coefficient (mem) | $D_{\text{AmB}}$ | $2 \times 10^{-12}$ | $\text{m}^2/\text{s}$ |
| Inter-membrane distance | $d_{\text{ch}}$ | 100 | μm |
| Membrane thickness | $d_{\text{mem}}$ | 125 | μm |
| Residence time | $t_{\text{res}}$ | 1 | s |
| Length of channel | $l_{\text{ch}}$ | 0.1 | m |
| Width of channel | $w_{\text{ch}}$ | 0.1 | m |
| Viscosity of water | $\mu$ | $9 \times 10^{-4}$ | Pa-s |
| Porosity | $\varepsilon$ | 0.8 | |
| Velocity of dilute sol. | $v_{\text{lc}}$ | 0.01 | m/s |
| Membrane lifetime | $t_{\text{mem}}$ | 4 | year |
| Plant lifetime | $t$ | 20 | year |
| Operational hours per year | $t_{\text{a}}$ | 8000 | h |
| Pump efficiency | $\eta_{\text{pmp}}$ | 0.75 | |
| Faradaic efficiency | $\eta_{\text{f}}$ | 0.95 | |
| Production capacity | $m_{\text{H2}}^{\text{cap}}$ | 1500 | kg/day |
| Pumping system cost | $c_{\text{pmp}}$ | 300 | €/kW |
| Membrane cost | $c_{\text{m}}$ | 150 | $€/\text{m}^2_{\text{mem}}$ |
| Waste heat price | $c_{\text{ht}}$ | 0.01 | €/kWh |
| Discount factor | $r$ | 0.06 | |
| Infrastructure price | $c_{\text{inf}}$ | 1123 | $€/\text{m}^2_{\text{space}}$ |
| Regeneration system price | $c_{\text{reg}}$ | 3400 | $€/\text{m}^2_{d\text{.sol}}/\text{h}$ |
| Labour cost | $c_{\text{lab}}$ | 20% | (% of CAPEX) |
| RED stack | $C_{\text{REDstack}}$ | $c_{\text{m}}$ | $€/\text{m}^2_{\text{mem}}$ |

$$E_{\text{u.c}}^{\text{oc}} = (a_{\text{m,aem}} + a_{\text{m,cem}}) \frac{RT}{F} \ln \left( \frac{\gamma_{\text{hc}} c_{\text{hc}}}{\gamma_{lc} c_{lc}} \right) \tag{8.1}$$

where $a$ is the permselectivity of IEMs measured at a concentration $c_{\text{hc}}$ and $c_{\text{lc}}$ at a constant temperature of 298 K for a specific membrane. Here we assume the same $a$ for both membranes. $F$ is the Faraday constant. $T$ is the room temperature. $R$ is the ideal gas constant. The activity coefficient of solutions ($\gamma$) is a measure of deviation from ideal activity due to energy lost in divesting an ion from a cloud of oppositely charged ions. The activity coefficients depend on molal salt concentration.

$$\ln(\gamma) = -|z_M z_X| f^\gamma + 2m \frac{\nu_M \nu_X}{\nu} B_{MX}^\gamma + 3m^2 \frac{(\nu_M \nu_X)^{3/2}}{\nu} C_{MX}^\gamma \tag{8.2}$$

$$\phi = 1 - |z_M z_M| A_\phi \left( \frac{I^{0.5}}{1 + bI^{0.5}} \right) + 2m \frac{\nu_M \nu_X}{\nu} B_{MX}^\gamma + 2m^2 \frac{(\nu_M \nu_X)^{3/2}}{\nu} C_{MX}^\gamma \tag{8.3}$$

where $m^0 = 1$ mol/kg is the standard molality, $b = 1.2\ \text{kg}^{1/2}\ \text{mol}^{-1/2}$ is a constant for all solutes; $a$ is a numerical constant equal to 2 for univalent ions; $I$ is the ionic strength; $A_\varnothing$, is the Debye–Huckel parameter for the osmotic coefficient; $z_M$ and $z_X$ are the charges of cation and anion; $B_{MX}^\gamma$ and $C_{MX}^\gamma$ are the ion-interaction parameters or virial coefficients.

The virial coefficients for $NH_4HCO_3$ solutions at 298 K are reported as follows. Ion-interaction parameters or virial coefficients are related to the short-range interaction between the ions, thus being particularly important when the salt concentration is high. The virial coefficients are a function of the electrolyte type, the temperature and the pressure. The correlations to evaluate the interaction parameters were found in literature as a function of temperature and pressure [25].

$$f^\gamma = -0.392 \left( \frac{I^{0.5}}{1 + bI^{0.5}} + \frac{2}{b} \ln\left(1 + bI^{0.5}\right) \right) \tag{8.4}$$

$$B_{MX}^\gamma = 2\beta_{MX}^0 + \frac{2\beta_{MX}^0}{\alpha^2 I} \left( 1 - e^{-\alpha I^{0.5}} \left( 1 + \alpha I^{0.5} - \frac{\alpha^2 I}{2} \right) \right) \tag{8.5}$$

here $C_{MX}^\gamma = \frac{3}{2} C_{MX}^\phi, \alpha = 2, b = 1.2, \beta^0 = -0.038, \beta^1 = 0.07$

At other concentrations, the $H_2$ production rate decreases either due to an increase in ohmic resistance or a decrease in concentration ratio. The waste heat required for restoring the concentrations was found to increase with the increase in dilute solution concentration.

There is an internal loss in the RED cell due to its components and operating parameters. This internal resistance consists of ohmic and non-ohmic resistances. The ohmic resistance for a unit cell ($\Omega m^2$) is the cumulative sum of membrane and channel (concentrate and dilute) resistances.

$$R_{u.c} = R_{lc} + R_{hc} + R_{m,aem} + R_{m,cem} \tag{8.6}$$

The channel ohmic resistance ($R_{ch}$) is the resistance ($\Omega m^2$) due to the conductivity of the solution in the channel and spacer geometry. It depends on concentration and is calculated using the molar conductivity of the salt.

$$R_{ch} = \frac{\delta_{ch}}{\varepsilon k_{ch} c_{ch}} \tag{8.7}$$

Here $\delta_{ch}$ is the intermembrane distance in μm, $\varepsilon$ is the porosity of spacers, $k_{lc}$ and $k_{hc}$ are the conductivities of the feed solutions. The conductivity of the feed solutions (mS/m) was calculated using the Jones–Dole equation at a constant temperature of 298 K.

$$k = \Lambda \cdot c_i \tag{8.8}$$

$$\Lambda = \Lambda_0 - \frac{A_\Lambda \Lambda_i c_i^{1/2}}{1 + B_\Lambda c_i^{1/2}} - C_\Lambda c_i \tag{8.9}$$

Here, $A_\Lambda$, $B_\Lambda$, $C_\Lambda$ are model parameters used for fitting and referred from Bevacqua et al., $c^i$ is the inlet concentration; $\Lambda_0$ is the molar equivalent conductivity of salt at infinite dilution [16]. The area-specific membrane resistance ($R_\mathrm{m}$) when immersed in the solution is estimated from the following expression,

$$R_\mathrm{m} = a \cdot (c_\mathrm{lc})^{-0.236} \tag{8.10}$$

This expression shows membrane resistance as a function of concentration. $R_\mathrm{m}$ is area-specific membrane resistance for cation and anion exchange membrane, $c^i_\mathrm{lc}$ is the inlet concentration of low concentrated solution, a is a fitting parameter ($\Omega\mathrm{m}^2$/M) [16]. The non-ohmic resistance is composed of the concentration change ($R_{\Delta\mathrm{c}}$) and the boundary layer thickness ($R_\mathrm{b.l}$). $R_{\Delta\mathrm{c}}$ is the resistance due to concentration change in the bulk solution along the length of the channel. $R_\mathrm{b.l}$ is the resistance from the concentration gradient perpendicular to the membrane surface within each compartment. The non-ohmic resistances are dependent on current density.

The actual unit cell potential ($E^\mathrm{act}_\mathrm{u.c}$) is the potential across the RED unit cell. The potential drops due to ohmic resistances in the RED unit cell,

$$E^\mathrm{act}_\mathrm{u.c} = E^\mathrm{oc}_\mathrm{u.c} - R_\mathrm{u.c}\, j^\mathrm{pp}_\mathrm{u.c} \tag{8.11}$$

Alkaline electrolysers operate above the reversible potential for water electrolysis of 1.23 V. This is due to overpotentials, at low current densities the electrode potentials contribute the most. The anodic and cathodic overpotentials are estimated to be 300 mV in total at low current densities. Hence, the required potential for water electrolysis is assumed 1.53 V [26]. In order to meet this potential, unit cells are stacked together in series. The minimum number of unit cells to be stacked in series is calculated as

$$N_\mathrm{u.c} = \frac{1.53}{E^\mathrm{oc}_\mathrm{u.c}} \tag{8.12}$$

The RED stack resistance ($R_\mathrm{s}$) is the total resistance, including $N$ unit cells and two electrodes in a stack.

$$R_\mathrm{s} = N\, R_\mathrm{u.c} + R_\mathrm{e} \tag{8.13}$$

Here $R_e$ ($\Omega m^2$) is the resistance from the electrodes in the electrode rinse solution. However, its contribution to the loss in performance is not significant for large stacks and hence neglected [27–29]. The current density ($j$) (A/m$^2$) is calculated using the following equation based on Ohm's law,

$$j_{\mathrm{u.c}}^{\mathrm{pp}} = \frac{E_{\mathrm{u.c}}^{\mathrm{oc}}}{2R_{\mathrm{u.c}}} \tag{8.14}$$

The stack open circuit potential ($E_{\mathrm{s}}^{\mathrm{act}}$) (V) drops upon connecting the RED device to an external load. Thus, the actual voltage across the terminals of the stack is given by

$$E_{\mathrm{s}}^{\mathrm{act}} = N\, E_{\mathrm{u.c}}^{\mathrm{act}} \tag{8.15}$$

The theoretical amount of $H_2$ produced per unit time in the compartment with an electrode–electrolyte rinse solution is the $H_2$ production rate ($\dot{n}_{H_2}$) and is calculated as

$$\dot{n}_{H_2} = \frac{j_{\mathrm{u.c}}^{\mathrm{pp}}\, 3600}{zF} \eta_{\mathrm{F}} \tag{8.16}$$

Here, $z = 2$ is the ion valence per mole of $H_2$ gas, $\eta_{\mathrm{F}}$ is the Faradaic efficiency. $\eta_{\mathrm{F}}$ signifies that not all the current generated is utilised to make $H_2$ due to undesired reactions or loss in the form of heat. In a RED system, the loss in Faradaic efficiency is due to back diffusion of ions, transport of co-ions and osmosis (i.e. closely related to membrane permselectivity), and ionic short-circuiting in the feed and drain channels. This loss in Faradaic efficiency can be as high as 50% in a typical closed-loop RED system in comparison with alkaline water electrolysers where the faradaic efficiency range from 5 to 25% due to other reasons [30–36]. Assuming the membrane near to the ideal, the Faradaic efficiency is assumed to be 0.95.

Membrane area ($A_{\mathrm{mem}}^{\mathrm{tot}}$) is the total membrane area (m$^2$) required to produce 1500 kg $H_2$ a day ($\dot{m}_{\mathrm{H2}}^{\mathrm{cap}} = 0.0174$ kg/s is the required production capacity).

$$A_{\mathrm{mem}}^{\mathrm{tot}} = \frac{I_{\mathrm{st.}}^{\mathrm{p}}\, N_{\mathrm{u.c}}}{j_{\mathrm{u.c}}^{\mathrm{pp}}} \tag{8.17}$$

$I_{\mathrm{st.}}^{\mathrm{p}}$ is total current from the stacks in parallel. It is calculated as

$$I_{\mathrm{pp}}^{\cdot\cdot} = \frac{\dot{m}_{H_2}^{\mathrm{cap}}\, zF}{2.016\, \eta_{\mathrm{F}}} 1000 \tag{8.18}$$

where $\dot{m}_{H_2}^{\mathrm{cap}}$ is the required $H_2$ production capacity (kg/s), and the molecular weight of $H_2$ is 2.016 g/mol.

## Pumping System Model

Experimentally, a pressure drop occurs over the inlet and outlet of the feed solutions. To compensate for the loss, pumping of the feed solutions is essential. In this section, the calculations of flow parameters such as fluid flow rate, pressure drop and pumping power are shown.

The flow velocity of the solution influences the non-ohmic resistance and thus the power density [37–39]. The flow rate ($Q_{ch}$) ($m^3/s$) for the feed solution is estimated as

$$Q_{ch} = v_{ch}\delta_{ch}w\varepsilon \tag{8.19}$$

Here, $\varepsilon$ is the porosity of the spacer, $v_{lc}$ is the velocity of the feed solution in the low concentrate channel (m/s), $\delta_{lc}$ is the intermembrane distance (m), $w$ is the width of the channel. The spacer filaments obstruct the flow through the channel, and this requires additional pumping power. To estimate the influence of spacer filaments, the hydraulic diameter ($d_{h,\ ch}$) of the spacer filled channel was determined using the following equation:

$$d_{h,ch} = \frac{4\varepsilon}{\frac{2}{\delta_{ch}} + \left((1-\varepsilon)\frac{8}{\delta_{ch}}\right)} \tag{8.20}$$

Assuming an ideal case of fully developed laminar flow, the pressure drop ($\Delta p_{ch}$) (Pa) was estimated using the Darcy–Weisbach equation,

$$\Delta p_{ch} = \frac{12\,\mu_{ch}\,l^2}{0.25\,d_h^2\,t_{res}} \tag{8.21}$$

where $\mu$ is the dynamic viscosity of water (Pa-s), $t_{res}$ is the residence time (s), $l$ is the length of the channel. The pumping power ($P_{pump}$) ($W/m^2$) is required to overcome the hydraulic resistance in pumping the feed solutions through the channels. This strongly depends on the spacer porosity, as it dictates the hydraulic radius, which influences the pressure drop. The Reynolds number (Re) indicates the dynamic fluid regime inside each compartment in a standard way. For a wide channel corrected for spacer porosity, Re is defined as [40, 41],

$$\mathrm{Re} = \frac{\rho_{ch}v_{ch}d_{h,ch}}{\mu_{ch}} \tag{8.22}$$

## Mass Balances

The concentration difference between the adjacent channels drives the salt flux from the high concentration to the low concentration channel. There are three species transported across an IEM, co-ions, counter-ions and water. The water diffuses through the IEM from dilute solution to concentrate solution due to osmotic pressure. Also, solvated ions take with them water molecules, which is called electroosmotic drag.

The total solute transport, i.e. salt flux ($J_{salt}$) across an IEM is the sum of coulombic and diffusive ion transport. The migrative flux ($J_{mig}$) is the salt transferred with the counter-ions migration through the IEM. The diffusive flux ($J_{dif}$) is the salt flux due to the co-ion's diffusion. The total salt flux is given as

$$J_{\text{salt}} = J_{\text{mig}} + J_{\text{diff}} = \frac{j}{F} + 2\frac{D_{\text{AmB}}\left(c_{\text{hc}}^{\text{in}} - c_{\text{lc}}^{\text{in}}\right)}{\delta_{\text{mem}}} \tag{8.23}$$

Here $\delta_{mem}$ is the thickness of the IEMs in operation (m), $D_{AmB}$ is the diffusivity coefficient of ammonium bicarbonate in the membrane ($m^2/s$). The first term relates to coulombic (counter-ion) transport, while the second term relates to co-ion transport. The factor 2 relates to the number of membranes in a unit cell. The water transport includes osmotic and electroosmotic effects. The osmotic pressure difference between the two solutions creates a water flux ($J_{osm}$) directed from the dilute to the concentrate channel. The counter ions drag water molecules in the solvation shell in a direction opposite to the osmotic flux. This flux is known as the electro-osmotic flux ($J_{ele.osm}$). The water transport across an IEM is,

$$J_{\text{H}_2\text{O}} = J_{\text{osm}} + J_{\text{eosm}} = -\frac{2D_{\text{w}}\left(c_{\text{c}}^{\text{in}} - c_{\text{d}}^{\text{in}}\right)}{\delta_{\text{mem}}} + t_{\text{w}}J_{\text{salt}} \tag{8.24}$$

here $t_w$ is moles of water carried by one mole of salt. The concentration changes along the channel length due to ions diffusing through an IEM. The outlet concentrations of the feed solutions from a channel,

$$c_{\text{hc}}^{\text{o}} = c_{\text{hc}}^{\text{i}} - \frac{J_{\text{salt}}A}{Q_{\text{ch}}} \tag{8.25}$$

here $Q_{ch}$ is the inlet volume flow rate ($m^3/s$).

## Waste Heat Regeneration System

The regeneration system restores the outlet concentration of the concentrate and dilutes solutions to the corresponding inlet concentrations. It includes stripping and an absorption processes. The air-stripping column decomposes the outlet solution from the dilute compartment to ammonia gas and $CO_2$ gas at 333 K. The absorption unit dissolves the decomposed gases at 293 K in the outlet of the concentrate channel.

**Table 8.4** Input values for parameters as a function of inlet solution concentration for stripping column

| Concentration range | $a_1$ | $a_2$ | $a_3$ | $a_4$ | $a_5$ | $a_6$ | $a_7$ |
|---|---|---|---|---|---|---|---|
| $0.025 \leq C_1 = 0.1$ | 12.115 | 0.26 | 261.787 | 0.615 | 297.46 | 0.252 | 0.478 |
| $0.1 < C_1 \leq 0.2$ | 12.836 | 1.02 | 258.324 | 0.612 | 260 | 0.165 | 0.517 |
| $0.2 < C_1 \leq 0.56$ | 13.195 | 0.686 | 60.592 | 0.667 | 55.934 | 0.687 | 0.212 |
| $0.56 < C_1 \leq 2$ | 8.714 | 0.225 | 35.796 | 0.656 | 45.56 | 0.758 | 0.045 |

The heat required for restoring the concentrations to their initial concentration is the regeneration heat ($q_{\mathrm{reg}}$). The total amount of thermal power required to strip $Q_{\mathrm{lc}}$ of ammonium bicarbonate salt from dilute solution channel is given by,

$$q_{\mathrm{reg}} = q_{\mathrm{AmB}} Q_{\mathrm{lc}} A_{\mathrm{mem}}^{\mathrm{tot}} \cdot 3600 \tag{8.26}$$

$q_{\mathrm{AmB}}$ is the specific thermal duty (kWh/m$^3$) required to decompose ammonium bicarbonate solution into its components $NH_{4,\ (g)}$ and $CO_{2,\ (g)}$. The value was estimated using the relation $q_{AmB} = a_1 e^{a_2} C_1 - a_3 C_2^{a_4} + a_5 C_1^{a_6} C_2^{a_7}$ from Bevacqua et al. for the inlet concentration to stripping column $C_1$ and the outlet concentration $C_2$ from the stripping column [16]. Here, $a_1$ to $a_7$ are fitting parameters that are a function of $C_1$ (Table 8.4).

## Economic Model

The levelised cost of $H_2$ (LCH) is a simple method to assess the economic feasibility and provides close to a realistic assessment of the proposed process. The results of the economic assessment of the different technologies presented in this work are easy to compare, transparent and easy to understand. It is assumed that the costs and benefits from 1 year to the next are constant. The inflation rate is not considered.

$$\mathrm{LCH} = \frac{\mathrm{CAPEX} + \sum_{i=1}^{i=t} \frac{\mathrm{OPEX} + C_{\mathrm{heat}}}{(1+r)^t} + C_{\mathrm{mem,re}}}{\sum_{i=1}^{i=t} \frac{m_{\mathrm{H_2}}}{(1+r)^t}} \tag{8.27}$$

CAPEX, OPEX is the capital and the operating expenses, respectively. $C_{\mathrm{heat}}$ is the cost associated with the waste heat. $C_{\mathrm{mem,re}}$ is the membrane replacement cost. $m_{\mathrm{H_2}}$ is the amount of $H_2$ produced per year. Here $t$ is the plant lifetime (years).

The capital cost (CAPEX) (i.e. the capital investment (€)) of the plant was evaluated using the capital costs including the expenses related to purchasing and installation (labour and infrastructure) of all systems such as the RED stack (including membranes), the pumping unit, and the regeneration unit. The labour cost includes civil work associated with site preparation, the process equipment building, and offsite services and is assumed to be 20% of the capital cost. The cost of the regeneration system ($C_r$) was normalised to the flow rate ($m^3/h$), while the cost of the pumping system was normalised to the pumping power (kW). The cost of the membranes was normalised per membrane area. The cost of the membrane was estimated as the sum of production cost and raw material cost.

Using an analytical approach to investigate the cost of Nafion membranes, Minke et al. suggested that for constant membrane thickness an increase in production rate by four orders of magnitude results in the total membrane cost decreasing by almost one order of magnitude concerning the initial cost. With a decrease in thickness by almost seven times and a further increase in membrane production rate by one order of magnitude, the total membrane cost reduces to one-tenth of the present raw material cost [42]. Assuming a learning rate similar to that of Nafion membranes of 25 μm thickness, a minimum membrane cost was estimated. The production rate required was assumed to be $60 \times 10^8$ $m^2$ per year as a RED application requires large membrane areas due to low membrane potential and power density. The cost of the RED stack ($C_{Rst}$) includes the cost of membranes ($C_{mem}$), electrodes ($C_{ele}$), gaskets ($C_{gas}$), spacers($C_{spa}$) and end plates ($C_{e.p}$). It is assumed that the sum of costs of the electrodes, gaskets, spacers and end plates is the same cost as that of the membranes. The real estate cost ($C_{infra}$) was calculated using the dimensions of the RED cell and stack, according to Daniilidis et al. [19]. In Norway, the real estate prices are low in rural areas, and many of the sources of the waste heat are in rural areas. Hence, real estate prices can be neglected. The pumping system cost ($C_p$) consists of two pumps. The cost associated with one pump is $c_{pump}$ (€/kW). The cost associated with the electrolyte rinse solution (ERS) pump is neglected as the pumping required is $N$ (number of unit cells in a stack) times lower than for the feed solution pump.

$$\text{CAPEX} = C_{Rst} + C_{BOP} + C_L + C_I \tag{8.28}$$

$$C_{Rst} = C_{mem} + C_{gas} + C_{spa} + C_{ele} + C_{e.p} \tag{8.29}$$

$$C_{BOP} = C_r + C_p \tag{8.30}$$

$$C_m = c_{mem} A_{mem}^{tot} \tag{8.31}$$

The total pumping cost is normalised per unit stack area,

$$C_p = P_{pump} A_{mem}^{tot} c_{pump} \cdot 10^{-3} \tag{8.32}$$

The specific investment cost for the regeneration system is $c_{reg}$ (€/$m^3$/h). The total cost of the regeneration system is estimated for the required flow rate $Q_{lc}$ from a RED stack.

$$C_r = c_{reg} Q_{lc} A_{mem}^{tot} 3600 \tag{8.33}$$

The operational and maintenance cost (OPEX) (€) includes labour maintenance, and service and repairs expenses and are assumed to be 2% of CAPEX. Direct and indirect operational costs such as chemicals, emissions to air, water supply, labour, taxes, administration and insurance are not included.

The ion exchange membranes have a limited lifetime, and therefore, these membranes have to be replaced at the end of their lifetime until plant lifetime is reached. The cost associated with membrane replacement ($C_{\text{mem, re}}$) is calculated as,

$$C_{\text{mem,re}} = \sum_{n=1}^{n=f} \frac{C_{\text{m}}}{(1+r)^{t_{\text{m}} \cdot n}} \tag{8.34}$$

where $c_{\text{m}}$ and $t_{\text{m}}$ are the membrane costs (€/m$^2$) and membrane lifetime (years). $f$ is the factor that denotes the number of times replacement takes place ($f = t/t_{\text{mem}}$). The regeneration system requires heat for producing $NH_{3,g}$ and $CO_{2,g}$ from the LC outlet solution. This heat required by the process can be sourced from various thermal energy-intensive industries (e.g. cement, paper, biomass and waste incineration industries). Though it is said to be waste heat, a heat exchange system is required to provide the heat at a specific temperature. Hence, a cost ($C_{\text{heat}}$) is associated with it. This cost is assumed to be as low as 0.01 €/kWh [43].

$$C_{\text{ht}} = c_{\text{ht}} q_{\text{reg}} t_{\text{a}} \tag{8.35}$$

$t_{\text{a}}$ is the total number of operational hours per year (h). The sum of revenue earned by selling produced $H_2$ and operational expenses incurred on an annual basis is

$$\text{Profit} = \left( \sum_{i=1}^{i=t} \frac{m_{H_2} c_{H_2}}{(1+r)^t} \right) - \left( \text{CAPEX} + \sum_{i=1}^{i=t} \frac{\text{OPEX} + C_{\text{heat}}}{(1+r)^t} + C_{\text{mem,re}} \right) \tag{8.36}$$

where $c_{H_2}$ is $H_2$ selling price (€/kg$_{H2}$) and assumed to be 3.59 €/kg$_{H2}$

## Scenario Study

The study includes two scenarios based on the cost of the membrane, the membrane lifetime and the membrane properties (permselectivity and membrane resistance). In all the scenarios for ammonium bicarbonate solution, the $H_2$ production rate was

**Table 8.5** Input parameters for ammonium bicarbonate-based reverse electrodialysis

| Parameter | Present | Future | Unit |
|---|---|---|---|
| Concentrate sol. ($c_c$) | 2.6 | 2.6 | M |
| Dilute sol. ($c_d$) | 0.05 | 0.099 | M |
| Permselectivity (A) | 0.8 | 0.95 | |
| Membrane life ($t_{\text{mem}}$) | 4 | 10 | Years |
| Membrane resistance ($r_{\text{mem}}$) | $4.1 \times 10^{-4}$ | $6.9 \times 10^{-4}$ | Ωm$^2$ [16] |
| Cost of regeneration system ($c_{\text{reg}}$) | 3400 | 3000 | €/kwh/m$^3$ |
| Cost of waste heat ($c_{\text{ht}}$) | 0.01 | 0.005 | €/kwh |

**Table 8.6** Input–output parameters for scenario study

| Scenario | Input | Output |
|---|---|---|
| Present | $C_{mem}$, $T_{mem}$ | $M_{h2}$, $q_{req}$, $l_{ch}$, profit |
| Future | Profit, $t_{mem}$, $l_{ch}$ | $M_{h2}$, $q_{req}$, $c_{mem}$ |

calculated from the corresponding input data, as shown in Table 8.5. In the present scenario, the cost and lifetime of the membranes were fixed inputs, and with these, the LCH was estimated. In the future scenario, the LCH was assumed to be 1.71 €/kg$_{H2}$ as a target set by US DOE for $H_2$ produced from renewable energy sources.

In both scenarios, the concentrations of feed solutions were optimised to minimise the LCH. That is to maximise the $H_2$ production and to minimise the waste heat required. Finally, in the future scenario, profits of 10% were assumed to be made on an annual basis, the membrane life was increased to 10 years, and the membrane cost was estimated accordingly. The cell geometry (channel length and width) and flow parameters such as the residence time were assumed to be the same for all scenarios. Table 8.6 summarises all the input–output parameters.

## Results and Discussion

In general, an increase in production rate increases the total $H_2$ produced in the plant throughout its lifetime, which decreases the LCH. An increase in production rate can be achieved by an increase in current density, which indirectly relates to concentration ratio. High concentration ratios are achieved with high solubility of salt under the assumption that the electrical conductivity of the solution and permselectivity of membrane remain the same. The production rate is a function of peak current density, which in turn depends on the Nernst potential. This Nernst potential relates to the ratios of concentration or activity coefficients which are salt properties (e.g. solubility). The salts that can produce $H_2$ at a production rate higher than that of an AmB RED system due to the synergetic effect of higher OCP, higher solubility limit and lower resistance, need to be investigated experimentally. However, this is beyond the scope of this work. $H_2$ production rate can be increased by improving membrane properties such as permselectivity, conductance and reducing thickness while maintaining the mechanical strength. Further, the use of an optimal combination of electrolytic solutions, electrode material and their geometric configuration can decrease the cell overpotential and increase the overall stack potential, which in turn will improve the $H_2$ production rate. In the following subsections, we discuss a few parameters that have a significant influence on the $H_2$ production rate and waste heat required for restoring the salt concentration.

### Feed Solution Concentration

Figure 8.3 shows the influence of the dilute solution concentration on the $H_2$ production rate and on the waste heat required to restore the salt concentrations to

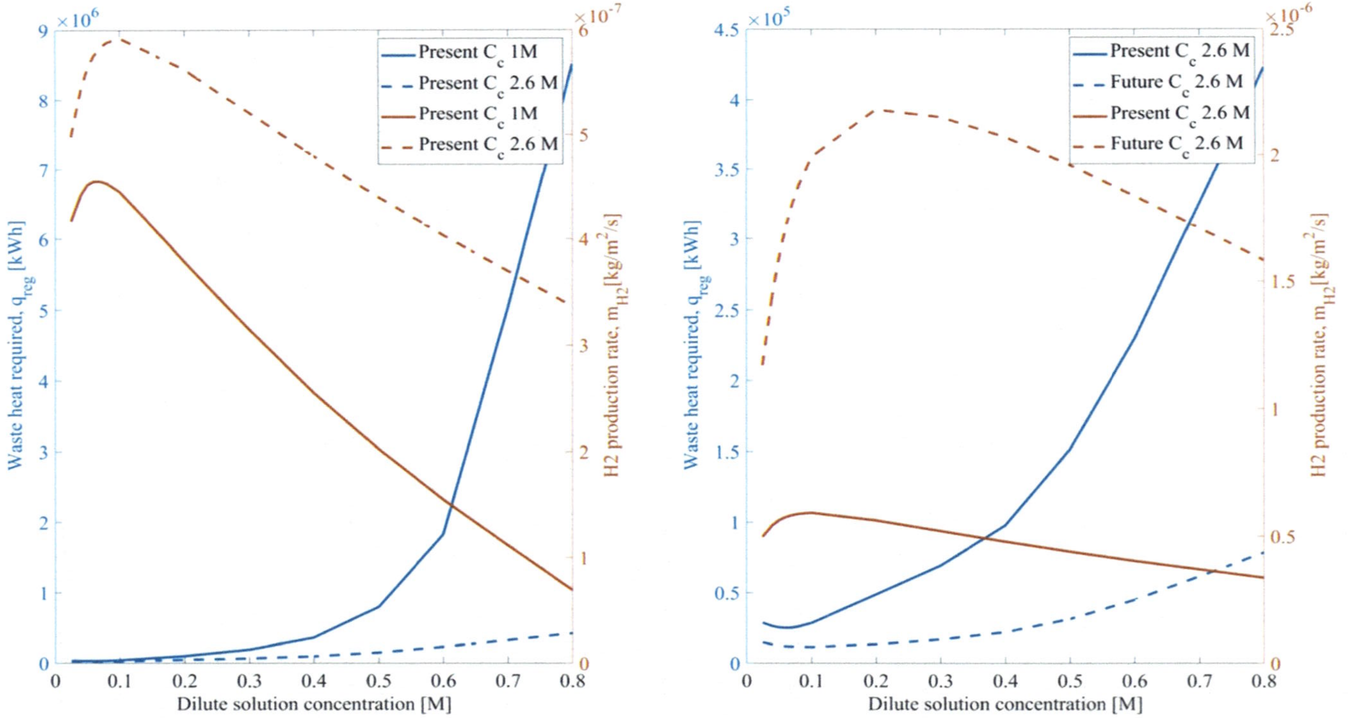


**Fig. 8.3** Waste heat required and $H_2$ production rate as a function of both dilute and concentrated feed solutions (left). Waste heat required and $H_2$ production rate as a function of both dilute feed solution for present and future scenario (right)

their initial values. At dilute solution concentrations below 0.1 M for a fixed concentration of concentrate, the concentration ratio ($c_c/c_d$) is high, but due to the high (ohmic) membrane and channel resistances, the electrochemical potential is low. The low electrochemical potential and high ohmic resistance result in a low current density. As the current density is directly proportional to the $H_2$ production rate, the latter decreases as well. The $H_2$ production rate increases until dilute solution concentrations 0.1 M and then decreases as shown in Fig. 8.3 (left). This decrease is due to a low concentration ratio ($c_c/c_d$). For the specified geometry and operating conditions of the RED cell, a maximum is observed at $c_c$ 2.6 M and $c_d$ 0.1 M, and 2.6 M and 0.2 M for present and future scenario, respectively, as shown in Fig. 8.3 (right).

For an increase in the concentration of concentrated solution the $H_2$ production increases. Also, the concentration of dilute solution at which the maximum $H_2$ production is observed increases with an increase in the concentration of concentrated solution.

An opposite trend is observed for the waste heat required for restoring the salt concentrations ($Q_{reg}$). For the specified geometry and operating conditions of the RED cell, an increase in the concentration of the dilute solution increases the waste heat required per unit volume to restore an increased amount of salt. At concentrations above $c_d = 0.07$ M and 0.1 M (for present and future scenario), the increase in dilute solution concentration increases the current density due to a decrease in (ohmic) the membrane and channel resistances. However, as the concentration ratio decreases the electrochemical potential decreases; this increases the number of cell pairs needed and thus increases the volumetric flow rate. The increase in volumetric flow rate causes an increase in the required waste heat.

Finding the optimum between the $H_2$ production rate and waste heat required is essential for a reduced LCH. This optimum is usually found at a concentration of dilute solution less than the concentration at maximum $H_2$ production as waste heat required dominates the LCH. When comparing the scenarios, as shown in Fig. 8.3 (right), the $H_2$ production rate increases by one order of magnitude and the waste heat required reduces by one order of magnitude. This is due to a high permselectivity and low membrane resistance. Further, the concentration of the dilute solution at which maximum $H_2$ production rate is observed shifts towards higher concentrations.

## Membrane Properties: Permselectivity and Membrane Resistance

Figure 8.4 shows the influence of the membrane resistance and permselectivity on the $H_2$ production rate and the waste heat required. For the specified geometry and operating conditions of the RED cell (fixed concentrations), an increase in membrane resistance decreases the electrochemical potential and thus, the peak current density decreases. This decrease in current density decreases the $H_2$ production rate. On the other hand, the increase in membrane resistance decreases current density which causes a decrease in the amount of salt transported through a membrane due to

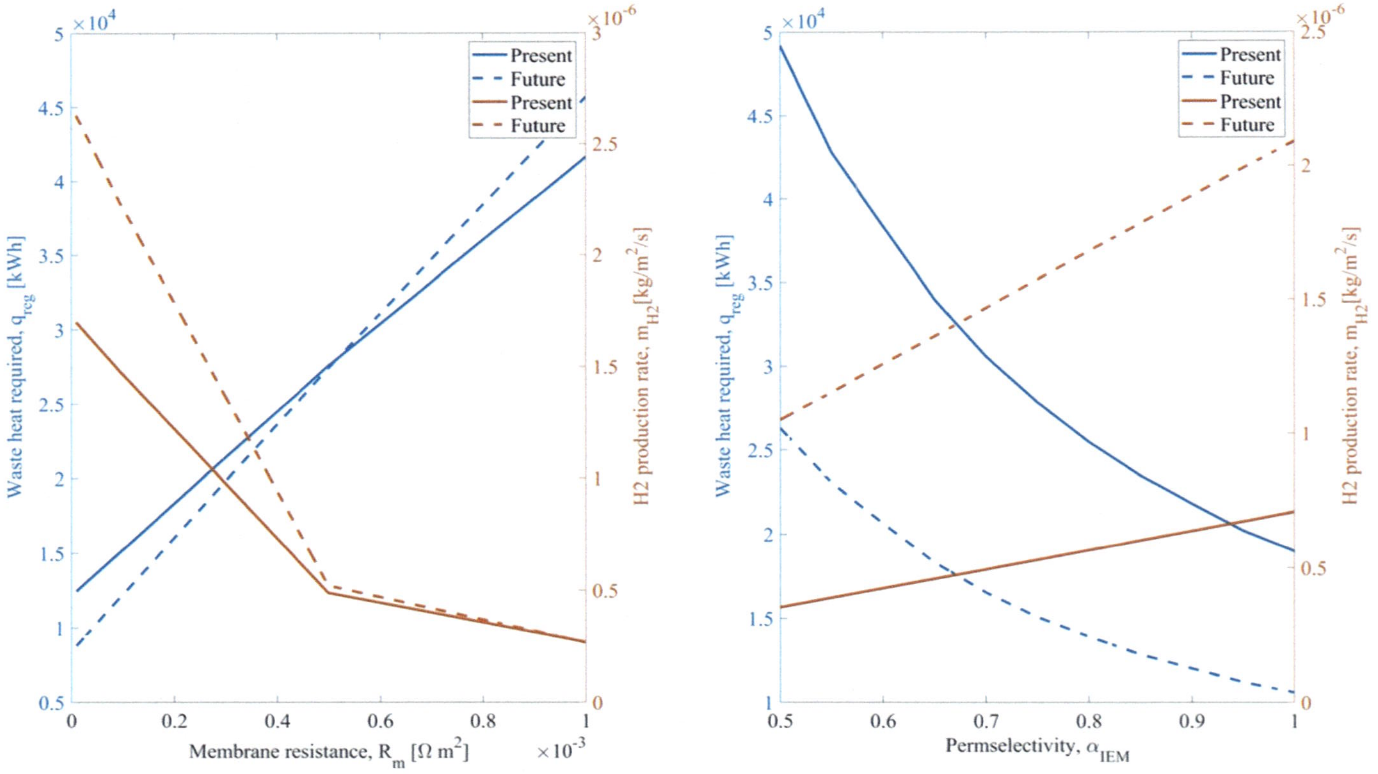


**Fig. 8.4** Influence of membrane resistance on waste heat required and $H_2$ production for present and future scenarios (left). Influence of permselectivity on waste heat required and $H_2$ production rate (right)

electromigration and decreases the amount of water dragged along due to electro-osmosis.

As the amount of salt flux is low, the outlet dilute solution concentration decreases. This decrease in dilute concentration causes a decrease in waste heat required per unit volume of a dilute solution ($q_{th}$ kWh/m$^3$). Nevertheless, the total (waste) heat required per kg of $H_2$ produced increases as the volume of the dilute solution at reduced current densities increases due to a higher number of cell pairs in a stack as shown in Fig. 8.4 (left). Membrane resistance is not only a material property; it is a ratio of the membrane thickness and its conductivity. With an increase in the membrane thickness, the length of the transport pathway increases; thus, the membrane resistance increases proportionally with the membrane thickness. There is a trade-off between membrane resistance and permselectivity. With the increase in feed solution concentrations, the permselectivity decreases due to a decrease in the charge-exclusion ability of membrane while increases the membrane conductivity due to an increase in ion concentration inside the membrane [44].

In the case of permselectivity, the increase in permselectivity increases the electrochemical potential. This increases the current density, which increases the $H_2$ production rate. As the electrochemical potential increases, the number of cell pairs decreases, which relates to a smaller dilute solution volume. This relates to less total heat required. With an increase in permselectivity, the optimum dilute solution concentration for low waste heat required decreases as shown in Fig. 8.4 (right).

## Cell Geometry: Residence Time and Channel Thickness

As shown in Fig. 8.5 (left), with an increase in residence time, the feed solution velocity and volume flow rate decreases (at $t_{res}$ 50s the feed solution flow velocity is $1.6 \times 10^{-3}$ m/s). This decreases the pressure drop across the inlet and the outlet of the RED stack. This results in a decrease in pumping power needed. Nevertheless, the increase in residence time increases the amount of salt diffusing through the membrane, which increases the waste heat per unit volume. However, as the volume flow rate decreases the total heat required to restore the concentrations decreases. As we operate at peak power current density, the influence of the resistance from the diffusion boundary layer and the electrical double layer on current density and electrochemical potential is neglected. Hence, the $H_2$ production rate is not affected by residence time. However, with an increase in residence time, these resistances will tend to dominate, and the $H_2$ production rate would decrease as well.

An increase in the channel thickness increases the ohmic resistance of the channels, which in turn decreases the actual unit cell potential. The decrease in open circuit potential decreases the peak power current density, which in turn decreases the $H_2$ production rate and the salt flux. Hence, the waste heat per unit volume of dilute solution required also decreases. However, the increase in channel thickness causes an increase in feed solution volume flowing through the channel. This causes an increase in the total waste heat required to regenerate the ammonium bicarbonate solutions. This increase in the heat requirement nullifies the decrease in

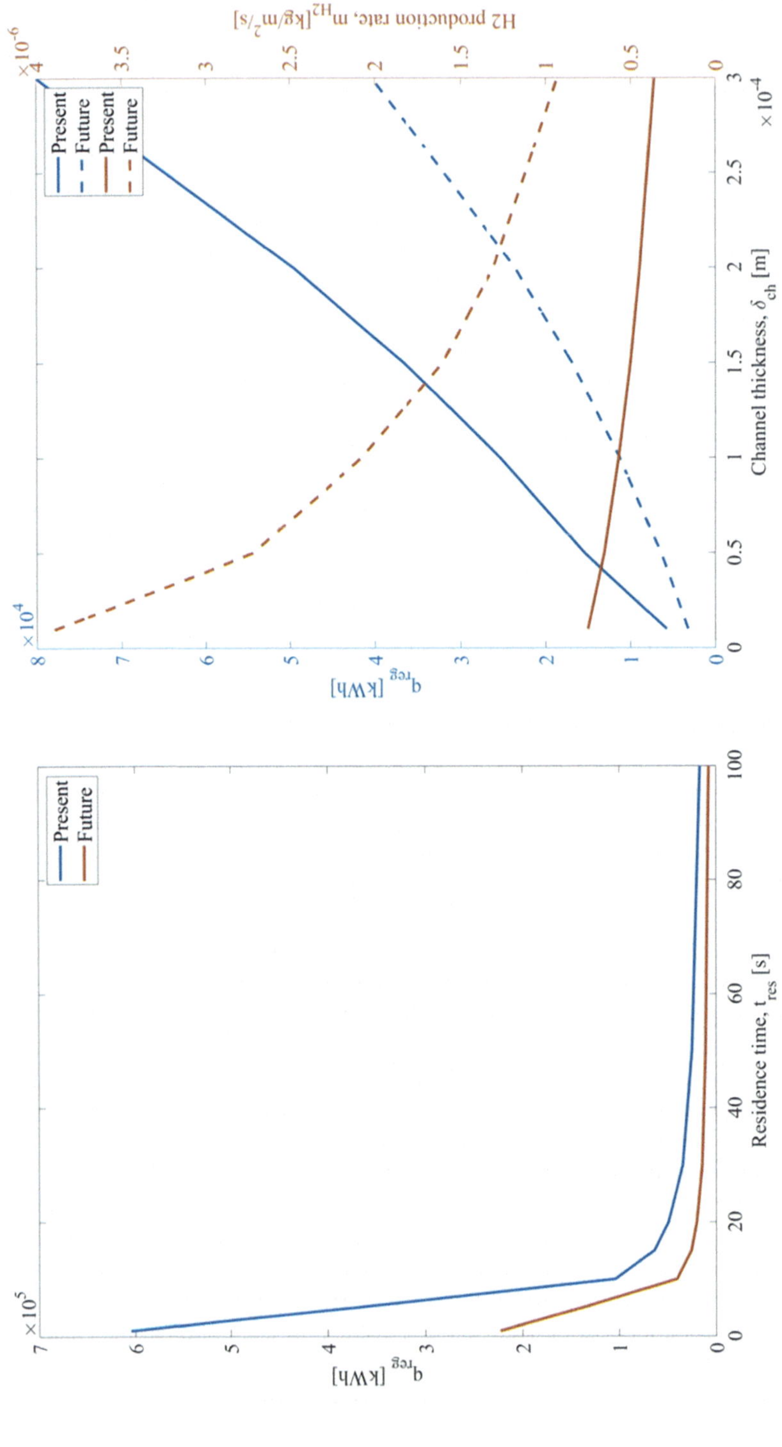


**Fig. 8.5** Waste heat required and $H_2$ production rate as a function of residence time (left) and channel thickness (right)

waste heat per unit volume due to the decrease in salt flux. Thus, the net effect observed is an increase in the total heat required, as shown in Fig. 8.5 (right). Moreover, the increase in channel thickness increases the feed solution flow rate. The increase in flow rate increases the amount of water per gram of salt transported; thus $J_{osm}$ increases. Hence, the total heat required to recover the same amount of salt increases.

## Economic Analysis: Membrane Cost, Lifetime and Waste Heat Cost

The membrane resistance and permselectivity have linear and non-linear relation with LCH, as shown in Fig. 8.6 (left). While the increase in membrane resistance increases LCH, increase in permselectivity decreases LCH. In case of future scenario, at permselectivity as high as 0.95 and membrane resistance as low as 0.1 $\Omega m^2$ the LCH can reach the US DOE target for other thermodynamic and economic parameters are kept constant. The membrane cost and the cost of waste heat have a linear relation with the LCH, as shown in Fig. 8.6.

For every euro, the membrane cost increases the LCH increases by 0.055 €/$kg_{H2}$ and 0.01 €/$kg_{H2}$ for present and future scenario, respectively. In case of the cost of waste heat, for every 0.001 €/kWh increase the LCH increases by 4.02 and 1.78 €/ $kg_{H2}$. Membrane life of more than 4 years has no significant influence on the LCH.

## Economic Comparison: Capex and LCH

Figure 8.7 shows a comparison of the present and future scenario in terms of cost composition. The Capex composition, in case of the present scenario, is dominated by 54.75% for the regeneration system and 28% for the membranes and RED stack. In the case of the future scenario, the regeneration system contributes the most with 76.53% as shown in Fig. 8.7 (left). The share of Capex for the regeneration system is high because of the high-volume flow rate of the dilute solution and a high amount of heat required per unit volume of the dilute solution.

In the case of LCH composition for the present scenario, the costs associated with waste heat and Capex contribute most with 41.72% and 36.52%, respectively, as shown in Fig. 8.7 (right). The major contributor to Capex is the regeneration system. Hence, the regeneration system must be optimised to reduce its contribution to LCH further and thus reducing LCH.

## Conclusion

A simplified thermodynamic RED model and an LCH model are developed for a thermally driven AmB RED system. The thermodynamic model includes various parameters such as the concentration of the feed solutions, membrane properties, channel geometry. The LCH model includes capital costs-associated RED stack and

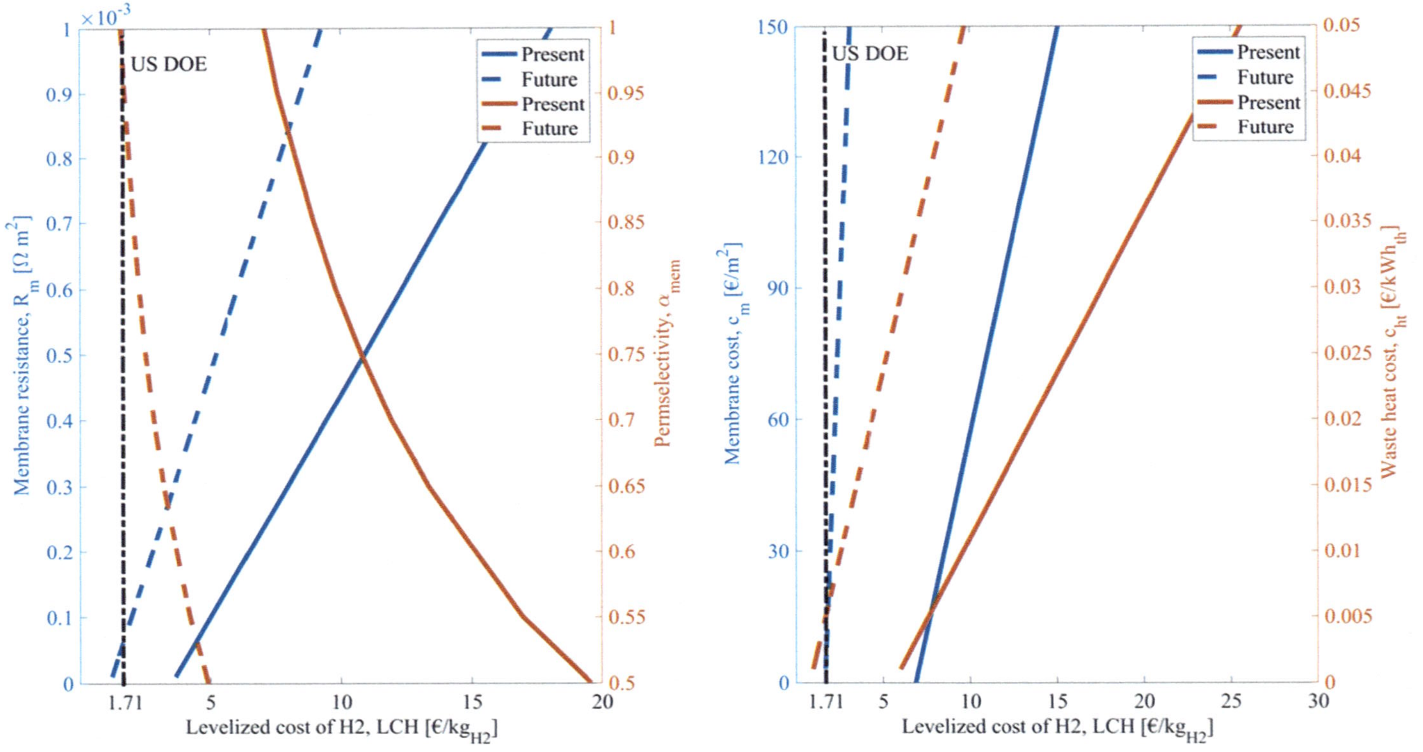


**Fig. 8.6** Influence of membrane resistance and permselectivity on levelised cost of $H_2$ (left) and influence of membrane cost and cost of waste heat on levelised cost of $H_2$ (right)

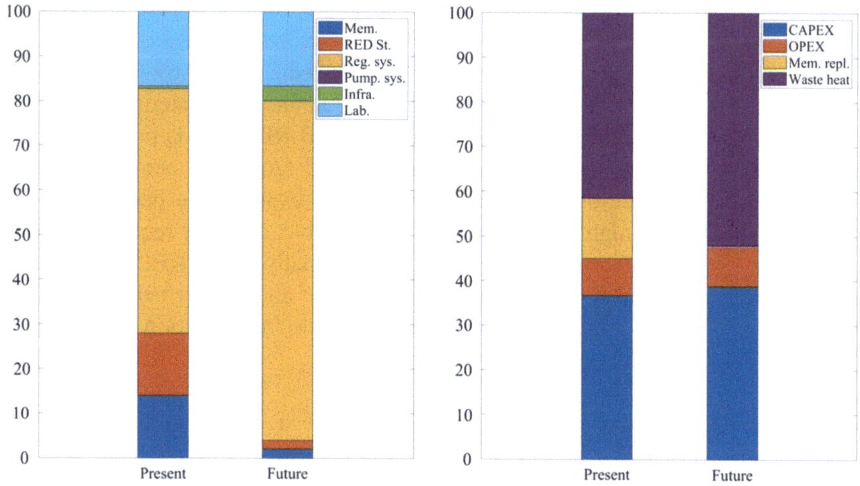


**Fig. 8.7** Comparison of component contribution to CAPEX for present and future scenario (left) and comparison of cost composition of LCH for present and future scenario (right)

regeneration system, and operational and membrane replacement costs of the AmB RED system. The LCH model also includes the cost associated with waste heat.

Maximum $H_2$ production is observed for dilute solution concentration of 0.099 M and 0.05 M for present and future scenario, respectively (at a concentrate solution concentration of 2.6 M).

Membrane conductance and permselectivity increase the $H_2$ production and decrease the waste heat required as they decrease the number of cell pairs and a decrease in the volume of the dilute solution entering the regeneration system.

Channel thickness increases channel resistance and thus, the $H_2$ production rate decreases. Also, the increase in channel thickness increases the total waste heat required as volume as well as the concentration of the dilute solution entering the regeneration system increases.

Increase in the residence time increases the waste heat per unit volume (due to increase in concentration) but decreases the total waste heat required (due to decrease in the volume of dilute solution entering regeneration system). Also, an increase in the residence time increases energy losses due to diffusion boundary layer and concentration change which decreases the $H_2$ production rate.

The scenario analysis includes two different cases, the present and the future. These scenarios differ in the inlet dilute solution concentration, the membrane properties, the cost of the membrane and the membrane lifetime. The scenario analysis demonstrates the cost of the membrane as the influential parameter in the present scenario; the cost of heating and the cost of the regeneration system, control the levelised cost of $H_2$ (LCH) in the future scenario. In both, the scenario, major contributor for Capex is regeneration system. Also, the cost of heating contributes the most to LCH, making optimisation of the regeneration system crucial to further minimise LCH.

In the future, the cost of the membrane can be reduced to 1.69 €/m$^2$ with an increase in production rate, a reduction in production and raw material cost, and a reduction in membrane thickness. It is clear that AmB RED for $H_2$ production has economic potential at membrane cost of less than 4.5 €/m$^2$, and membrane life of 4 years or more. Future efforts must be directed to evaluate the performance of potential salts that are highly conductive, highly soluble, and may require less heat for recovery, such as lithium bromide. Finally, a thermally driven RED system with a regeneration system cost of less than 3000 €/kWh/m$^3$ and a waste heat cost of less than 0.005 €/kWh can present an economically feasible for utilisation of low-grade waste heat for $H_2$ production.

**Acknowledgements** The authors are grateful to the ENERSENSE programme and NTNU Team Hydrogen at the Norwegian University of Science and Technology (NTNU) for supporting and helping on this book project.

## References

1. Miró L, Brückner S, Cabeza LF (2015) Mapping and discussing industrial waste heat (IWH) potentials for different countries. Renew Sust Energ Rev 51:847–855
2. Papapetrou M, Kosmadakis G, Cipollina A, La Commare U, Micale G (2018) Industrial waste heat: estimation of the technically available resource in the EU per industrial sector, temperature level and country. Appl Therm Eng 138:207–216
3. Law R, Harvey A, Reay D (2013) Opportunities for low-grade heat recovery in the UK food processing industry. Appl Therm Eng 53(2):188–196
4. Kim T, Rahimi M, Logan BE, Gorski CA (2016) Harvesting energy from salinity differences using battery electrodes in a concentration flow cell. Environ Sci Technol 50(17):9791–9797
5. Tamburini A, Tedesco M, Cipollina A, Micale G, Ciofalo M, Papapetrou M et al (2017) Reverse electrodialysis heat engine for sustainable power production. Appl Energy 206:1334–1353
6. Vermaas DA, Kunteng D, Saakes M, Nijmeijer K (2013) Fouling in reverse electrodialysis under natural conditions. Water Res 47(3):1289–1298
7. Luo X, Cao X, Mo Y, Xiao K, Zhang X, Liang P et al (2012) Power generation by coupling reverse electrodialysis and ammonium bicarbonate: implication for recovery of waste heat. Electrochem Commun 19(1):25–28
8. Hatzell MC, Logan BE (2013) Evaluation of flow fields on bubble removal and system performance in an ammonium bicarbonate reverse electrodialysis stack. J Memb Sci 446:449–455
9. Zhu X, He W, Logan BE (2015) Influence of solution concentration and salt types on the performance of reverse electrodialysis cells. J Memb Sci 494:154–160
10. Kwon K, Park BH, Kim DH, Kim D (2015) Parametric study of reverse electrodialysis using ammonium bicarbonate solution for low-grade waste heat recovery. Energy Convers Manag 103:104–110
11. Bevacqua M, Carubia A, Cipollina A, Tamburini A, Tedesco M, Micale G (2016) Performance of a RED system with ammonium hydrogen carbonate solutions. Desalin Water Treat 57 (48–49):23007–23018

12. Hatzell MC, Ivanov ID, Cusick R, Zhu X, Logan BE (2014) Comparison of hydrogen production and electrical power generation for energy capture in closed-loop ammonium bicarbonate reverse electrodialysis systems. Phys Chem Chem Phys 16(4):1632–1638
13. Nazemi M, Zhang J, Hatzell MC (2017) Harvesting natural salinity gradient energy for hydrogen production through reverse electrodialysis power generation. J Electrochem Energy Convers Storage 14(2):020702
14. Chen X, Jiang C, Zhang Y, Wang Y, Xu T (2017) Storable hydrogen production by reverse electro-electrodialysis (REED). J Memb Sci 544:397–405
15. Kim DH, Park BH, Kwon K, Li L, Kim D (2017) Modeling of power generation with thermolytic reverse electrodialysis for low-grade waste heat recovery. Appl Energy 189:201–210
16. Bevacqua M, Tamburini A, Papapetrou M, Cipollina A, Micale G, Piacentino A (2017) Reverse electrodialysis with NH4HCO3-water systems for heat-to-power conversion. Energy 137:1293–1307
17. Turek M, Bandura B (2007 Feb) Renewable energy by reverse electrodialysis. Desalination 205 (1–3):67–74
18. Post JW, Goeting CH, Valk J, Goinga S, Veerman J, Hamelers HVM et al (2010) Towards implementation of reverse electrodialysis for power generation from salinity gradients. Desalin Water Treat 16(1–3):182–193
19. Daniilidis A, Herber R, Vermaas DA (2014) Upscale potential and financial feasibility of a reverse electrodialysis power plant. Appl Energy 119:257–265
20. Weiner AM, McGovern RK, Lienhard VJH (2015) A new reverse electrodialysis design strategy which significantly reduces the levelized cost of electricity. J Memb Sci 493:605–614
21. Micari M, Cipollina A, Giacalone F, Kosmadakis G, Papapetrou M, Zaragoza G et al (2019) Towards the first proof of the concept of a reverse electrodialysis – membrane distillation heat engine. Desalination 453:77–88
22. Giacalone F, Papapetrou M, Kosmadakis G, Tamburini A, Micale G, Cipollina A (2019) Application of reverse electrodialysis to site-specific types of saline solutions: a techno-economic assessment. Energy 181:532–547
23. Papapetrou M, Kosmadakis G, Giacalone F, Ortega-Delgado B, Cipollina A, Tamburini A et al (2019) Evaluation of the economic and environmental performance of low-temperature heat to power conversion using a reverse electrodialysis – multi-effect distillation system. Energies 12 (17):3206
24. Raka YD, Karoliussen H, Lien KM, Burheim OS (2019) Opportunities and challenges for thermally driven hydrogen production using reverse electrodialysis system. J Hydrog Energy 45:1212–1225
25. Kim YP, Seinfeld JH (1995) Atmospheric gas-aerosol equilibrium: IV. Thermodynamics of carbonates. Aerosol Sci Technol 23(2):131–154
26. Burheim OS (2017) Electrochemical energy storage. In: Engineering energy storage. Academic, Cambridge, pp 1–76
27. Daniilidis A, Vermaas DA, Herber R, Nijmeijer K (2014) Experimentally obtainable energy from mixing river water, seawater or brines with reverse electrodialysis. Renew Energy 64:123–131
28. Post JW, Hamelers HVM, Buisman CJN (2008) Energy recovery from controlled mixing salt and fresh water with a reverse electrodialysis system. Environ Sci Technol 42(15):5785–5790
29. Długołęcki P, Gambier A, Nijmeijer K, Wessling M (2009) Practical potential of reverse electrodialysis as process for sustainable energy generation. Environ Sci Technol 43 (17):6888–6894
30. Veerman J, Post JW, Saakes M, Metz SJ, Harmsen GJ (2008) Reducing power losses caused by ionic shortcut currents in reverse electrodialysis stacks by a validated model. J Memb Sci 310 (1–2):418–430
31. Millet P, Grigoriev S (2013) Water electrolysis technologies. Renew Hydrog Technol Prod Purif Storage Appl Saf 19–41

32. Kingsbury RS, Coronell O (2017) Osmotic ballasts enhance faradaic efficiency in closed-loop, membrane-based energy systems. Environ Sci Technol 51(3):1910–1917
33. Kingsbury RS, Chu K, Coronell O (2015) Energy storage by reversible electrodialysis: the concentration battery. J Memb Sci 495:502–516
34. Moya AA (2017) Numerical simulation of ionic transport processes through bilayer ion-exchange membranes in reverse electrodialysis stacks. J Memb Sci 524:400–408
35. Mei Y, Tang CY (2017) Co-locating reverse electrodialysis with reverse osmosis desalination: synergies and implications. J Memb Sci 539:305–312
36. Tedesco M, Hamelers HVM, Biesheuvel PM (2016) Nernst-Planck transport theory for (reverse) electrodialysis: I. Effect of co-ion transport through the membranes. J Memb Sci 510:370–381
37. Zhu X, He W, Logan BE (2015) Reducing pumping energy by using different flow rates of high and low concentration solutions in reverse electrodialysis cells. J Memb Sci 486:215–221
38. Weiner AM, McGovern RK, Lienhard VJH (2015) Increasing the power density and reducing the levelized cost of electricity of a reverse electrodialysis stack through blending. Desalination 369:140–148
39. Długołęcki P, Ogonowski P, Metz SJ, Saakes M, Nijmeijer K, Wessling M (2010) On the resistances of membrane, diffusion boundary layer and double layer in ion exchange membrane transport. J Memb Sci 349(1–2):369–379
40. Vermaas DA, Saakes M, Nijmeijer K (2014) Enhanced mixing in the diffusive boundary layer for energy generation in reverse electrodialysis. J Memb Sci 453:312–319
41. Vermaas DA, Saakes M, Nijmeijer K (2011) Power generation using profiled membranes in reverse electrodialysis. J Memb Sci 385–386(1):234–242
42. Minke C, Turek T (2015) Economics of vanadium redox flow battery membranes. J Power Sources 286:247–257
43. Naik-Dhungel N (2012) Waste heat to power systems. US Environmental Protection Agency
44. Fan H, Yip NY (2018) Elucidating conductivity-permselectivity tradeoffs in electrodialysis and reverse electrodialysis by structure-property analysis of ion-exchange membranes. J Memb Sci 573:668–681

# 9 Liquid Air Energy Storage: Analysis and Prospects

Zhongxuan Liu, Federico Ustolin, Lena Spitthoff, Jacob J. Lamb, Truls Gundersen, Bruno G. Pollet, and Odne S. Burheim

Z. Liu
Department of Energy and Process Engineering, Norwegian University of Science and Technology, Trondheim, Norway

Department of Language and Literature, Norwegian University of Science and Technology, Trondheim, Norway
e-mail: zhongxuan.liu@ntnu.no

F. Ustolin
Department of Mechanical and Industrial Engineering, Norwegian University of Science and Technology, Trondheim, Norway
e-mail: federico.ustolin@ntnu.no

L. Spitthoff
Department of Energy and Process Engineering, ENERSENSE, Norwegian University of Science and Technology, Trondheim, Norway
e-mail: lena.spitthoff@ntnu.no

J. J. Lamb (✉)
Department of Electronic Systems, ENERSENSE, Norwegian University of Science and Technology, Trondheim, Norway

Department of Energy and Process Engineering, ENERSENSE, Norwegian University of Science and Technology, Trondheim, Norway
e-mail: jacob.j.lamb@ntnu.no

T. Gundersen
Department of Energy and Process Engineering, Norwegian University of Science and Technology, Trondheim, Norway
e-mail: truls.gundersen@ntnu.no

B. G. Pollet · O. S. Burheim
Department of Energy and Process Engineering, ENERSENSE and NTNU Team Hydrogen, Norwegian University of Science and Technology, Trondheim, Norway
e-mail: bruno.g.pollet@ntnu.no; odne.s.burheim@ntnu.no



J. J. Lamb, B. G. Pollet (eds.), *Micro-Optics and Energy*,
https://doi.org/10.1007/978-3-030-43676-6_9

## Introduction

The security of energy supply has always been a core item on the European political agenda. In 2006, it was listed as one of the cornerstones of the standard energy policy, alongside environmental objectives and economic competitiveness [1]. Currently, we mainly depend on a fossil-based economy. More than 80% of the energy consumption comes from (traditional) fossil fuels, which drives our economy and supports the industrial, transportation and buildings sectors in our society. However, there is a threat to climate change caused by emissions from fossil fuels, which means that we need to take actions [2]. One measure is to switch our energy focus from traditional non-renewable energy structures to renewable energy forms, and also increase efforts in waste-to-energy systems.

With the development of the technologies, renewable energy and power sources are getting cheaper and more competitive with existing ones (coal, oil, gas and nuclear). Once installed, these power sources have limited operational costs. This has led to a growing share of renewable energy sources. If the process industries are to become more energy efficient and sustainable, industrial energy systems need to integrate different energy sources such as renewable and waste-to-energy systems. However, it is essential to consider a critical barrier for the switch, which is the intermittent nature of renewable energy sources. This means that renewable power might hold the limited potential to substitute fossil fuels due to the requirement of stability in a sustainable energy market.

An exciting trend for industrial systems that is likely to be realised in the next decades is the use of distributed energy systems that are adopted systematically and holistically. A distributed energy system is an efficient, reliable and environmentally friendly alternative to the traditional energy system. With this system, energy conversion units are situated close to energy consumers, and large units are substituted by smaller ones [3]. The advantages of utilising distributed systems are mainly based on flexibility, location and networking. The flexibility of distributed energy systems is associated with their scalability and ability to utilise various energy conversion technologies and energy sources. An improvement can also be seen in the reliability of energy supply because of the tendency of distributed systems to reduce the vulnerability of the overall system. This is related to their ability to operate in networks and utilise local resources. Also, distributed energy systems are environmentally friendly because of the absence of large power plants and transmission lines [4].

Following the trends of the energy system means that renewable energy is used as a kind of energy source in the distributed energy system, and it seems to be the most promising future energy system. However, once renewable energy is introduced to industrial systems, the most critical consideration should be the stability and sustainability of the energy supply because of the intermittency of renewable energy sources. Therefore, it is necessary to adopt energy storage technologies to smoothen variations in supply and demand and guarantee supply during energy deficit periods.

There are many energy storage technologies already reviewed in the literature [5–7]. These technologies are currently at different levels of maturity with a few already proven for commercial-scale applications. These technologies can be broadly classified according to the purpose for which the energy is stored, and they include both thermal and electrical energy storage.

## Thermal Energy Storage (TES)

TES is one of the most widely used forms of energy storage. The TES principle is the same for all technologies: energy is supplied during off-peak periods; it is collected and stored in the form of heat (specific, latent or reaction heat), and later in peak periods it is recovered for use. According to the actual application, these processes may coincide or more than once in the storage cycle. There are three thermal energy storage methods, and these will be briefly described in the below paragraphs.

## Sensible Heat Energy Storage (SHES)

In this form, the energy storage is based on the specific heat of the material, which means that the material does not undergo any form of phase change within the temperature range required for the storage application [8]. The most common materials for high-temperature TES include concrete, cast ceramics and molten salts. Molten salts with high storage temperature have been applied in solar thermal technology; their main disadvantage is that most of them have high melting points and therefore, parasitic heating is required to keep them in the liquid phase, which can lead to additional energy consumption.

## Latent Heat Energy Storage (LHES)

In this case, the energy is stored through the phase change of the material. Such materials have a high potential for thermal energy storage compared to the non-phase changing counterparts due to the high latent heat associated with the phase change. For energy storage applications, the phase of the material changes (usually from solid to liquid) at a temperature matching the thermal input source [9].

## Thermochemical Energy Storage (TCES)

This technology stores heat through reversible reactions. During off-peak periods, surplus thermal energy is used to dissociate a chemical reactant into products in an endothermic reaction. The products are stored separately, pending periods when energy is needed. When energy is demanded, the stored products are mixed and will react to form the initial reactant in a reaction which is exothermic [10].

Table 9.1 shows typical characteristics for different thermal energy storage technologies, which can give guidance to the choice of proper TES technology. The efficiency $\eta$ for different technologies can be calculated as follows [11]:

**Table 9.1** Typical characteristics for thermal energy storage technologies [12]

| Characteristics | SHES | LHES | TCES |
|---|---|---|---|
| Capacity (kWh/ton) | 10–50 | 50–150 | 120–250 |
| Thermal power (MW) | 0.001–10 | 0.001–1 | 0.01–1 |
| Efficiency (%) | 50–90 | 75–90 | 75–100 |
| Temperature range (°C) | 30–100 (water) | 40–120 | 80–1000 |
| Storage duration | Day/month | Hour/month | Hour/day |
| Barriers | Low capacity per weight | High cost and material stability | High cost and complexity |

$$\eta = \frac{\text{energy recovered}}{\text{energy input} + \text{initial energy for storage}} \quad (9.1)$$

However, the choice depends mainly on operating temperature range, storage capacity and duration required.

## Electrical Energy Storage (EES)

Electrical energy storage is regarded as one of the most readily available forms of storing energy. Electricity in its form cannot be stored (except in superconductors at cold temperatures). The only way to store electricity is to convert it into a more stable energy form, which would be transformed back to electricity when needed. Various technologies can be used to convert electricity to other forms of energy that can easily be stored. These technologies are referred to as electrical energy storage technologies and can be grouped as follows:

- Mechanical energy storage
- Chemical energy storage
- Electrochemical energy storage (supercapacitors, electrolysis/FC or batteries)
- Superconducting magnetic energy storage
- Thermal energy storage [6]

In this section, not all EES technologies will be presented; only some mature technologies will be listed, including pumped hydroelectric energy storage (PHES, which belongs to mechanical energy storage), battery energy storage (which belongs to electrochemical energy storage), compressed air energy storage (CAES, which belongs to thermo-mechanical energy storage), and $H_2$ energy storage (HES, which belongs to chemical energy storage). Last but not least, liquid air energy storage (LAES) will be introduced.

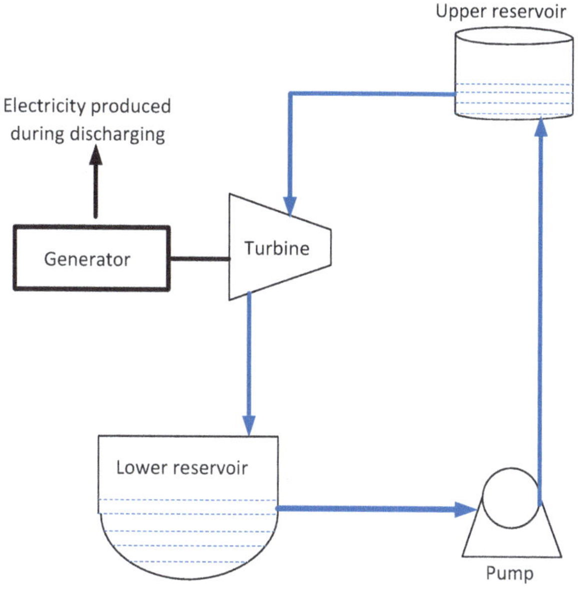


**Fig. 9.1** Schematic of PHES system

## Pumped Hydroelectric Energy Storage (PHES)

PHES is the most mature and widely used large-scale energy storage technology. Figure 9.1 shows the process of a PHES system that uses gravity to store energy. It stores electrical energy by pumping the water to a higher reservoir during off-peak periods when the energy is available. During peak periods, the water flows down to a lower reservoir, through a turbine where electricity is produced by a generator. The efficiency of the system is in the range of 65–85%. One limitation of the PHES is that several natural geological features are needed, and this is typically measured by a performance parameter that is defined as the adequate nearby land area divided by the adequate elevation [13].

## Battery Energy Storage (BES)

Battery technology is the most widespread energy storage device for power system applications, at least in terms of a number of devices (e.g. cellular phones, tablets and computers). The electricity is stored as chemical energy in a battery. It is an electrochemical device with the ability to deliver energy by converting chemical energy into electricity through electrochemical reactions. The round-trip energy storage efficiency is in the range of 60–95% depending on the operational cycle and electrochemistry. Batteries have short life cycles (compared to PHES); however, they have a moderate to decent efficiency and are quite cheap when the storage demand is small [13]. Sodium sulphur (NaS) is an example of batteries used for commercial energy storage in distribution grids for electricity. Its potential comes from the ability to provide high energy density (100–240 Wh/kg), excellent energy efficiency (>85%) and long discharge periods (approximately 7 h) [6]. Lead-acid ($PbO_2$) battery is the oldest rechargeable alternative for both commercial and household applications. It has a rated voltage of 2 V, energy and power density of about 30 Wh/kg and 150 W/kg, respectively. Its energy efficiency ranges from 85 to 90% with low maintenance and investment costs. Lithium-ion (Li-ion) batteries are used in a wide range of applications, such as portable electronics, medical devices, transportation and grid supports. Their energy and power density range from 90 to 190 Wh/kg and 500 to 2000 W/kg. They also have a high-efficiency range from 90 to 94% and average discharge times (about 2–4 h) [6].

## Compressed Air Energy Storage (CAES)

CAES systems use off-peak electricity to compress air and store it in a reservoir, either an underground cavern or above ground in pipes or vessels [14]. The air is released during peak periods, heated, expanded and used in a turbine and generator to produce electricity, as illustrated in Fig. 9.2. The CAES has an estimated efficiency in the range of 46–70% depending on the technology limit for the system, with an expected lifetime of about 40 years. The drawback of CAES is that it is geographically constrained, because the compressed air is preferably kept in underground caverns. Otherwise high-pressure tanks must be used to store the air above ground.

## Hydrogen Energy Storage (HES)

HES is one of the most promising chemical energy storages [15]. It has a high energy density. During charging, off-peak electricity is used to electrolyse water to produce $H_2$. The $H_2$ can be stored in different forms, e.g. compressed $H_2$, liquid $H_2$, metal hydrides or carbon nanostructures [16], which depend on the characteristics of available technologies. The stored $H_2$ is used in fuel cells or burned when electricity is needed. The best efficiency for the electrolyser is 70%, and the efficiency for the fuel cell is 50%.

The four abovementioned technologies (PHES, BES, CAES and HES) have been available for decades and will be described in more detail below.

## Liquid Air Energy Storage (LAES)

LAES has the advantage that this technology can be improved by using ideas from established technologies, such as gas liquefaction processes and air separation. The most important reason is that it is not geographically constrained, as the case is with, e.g. PHES. LAES consists of three distinct processes: charging, storing and discharging. During charging, excess electricity (e.g. from wind turbines) can

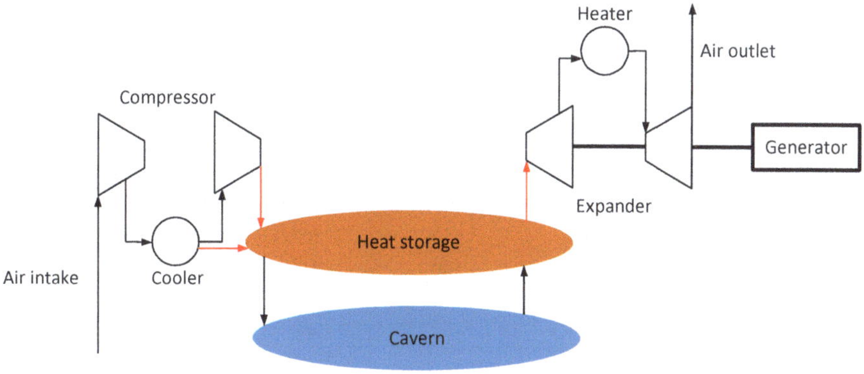


**Fig. 9.2** Schematic of an adiabatic CAES system

drive an air liquefaction process. Air from the environment is compressed in stages and then expanded to ambient pressure and sub-ambient temperature to generate the necessary refrigeration capacity to liquefy air. Liquid air is then stored in cryogenic tanks at nearly ambient pressure. During discharge, liquid air is pumped to high pressure, and then regasified and expanded through turbomachines to generate electricity and thereby recover the stored energy.

The charging process, which in essence is a liquefaction process, has direct influences on the liquid yield, and in turn the round-trip efficiency. Therefore, the liquefaction process is a crucial step in LAES processes. Four LAES technologies with different liquefaction processes are described and compared here, and a hybrid arrangement with higher system efficiency and performance is mentioned later.

## LAES Technologies

### Simulation of the Process Concepts

The simulation of four LAES technologies with different liquefaction processes was carried out using Aspen HYSYS. These processes are Linde–Hampson, Claude, Kapitza and a modified Claude process. The design parameters for the simulations are given in Table 9.2.

### Linde–Hampson Process

Linde proposed the simplest approach to liquefy air, and the cycle was patented in 1903 [17]. The Linde–Hampson process for liquefying air is shown in Fig. 9.3a. First, the air enters the compressor and is compressed to high pressure. Then the air goes through a cooler and a heat exchanger to be further cooled down. The

**Table 9.2** Design parameters for the processes

| Parameter | Liquefaction process | | | |
|---|---|---|---|---|
| | Linde–Hampson | Claude | Kapitza | Modified Claude |
| Ambient temperature (°C) | 25 | 25 | 25 | 25 |
| Ambient pressure (kPa) | 100 | 100 | 100 | 100 |
| Pressure after compressor (kPa) | 20,000 | 6000 | 6000 | 18,000 |
| Pressure after pump (kPa) | 6500 | 6500 | 6500 | 6500 |
| Liquid air storage pressure (kPa) | 100 | 100 | 100 | 100 |
| Liquid air storage temperature (°C) | –194.3 | –194.3 | –194.3 | –194.3 |
| Propane minimum temperature (K) | – | – | – | –180 |
| Propane maximum temperature (K) | – | – | – | –59 |
| Methanol minimum temperature (K) | – | – | – | –59 |
| Methanol maximum temperature (K) | – | – | – | 15 |
| Minimum approach temperature in cold box HX (K) | 5 | 5 | 5 | 5 |
| Minimum approach temperature in intercoolers (K) | – | – | – | 10 |
| Heat exchanger relative pressure loss | 0 | 0 | 0 | 0 |
| Isentropic efficiency of air turbines (%) | 85 | 85 | 85 | 85 |
| Isentropic efficiency of air compressors (%) | 85 | 85 | 85 | 85 |
| Isentropic efficiency of cryoturbine (%) | 70 | 70 | 70 | 70 |
| Recirculation ratio (%) | 7 | 80 | 80 | 15 |

recirculating vapour removes the heat of the air from the separator. The Joule–Thompson (J–T) valve expands the air to ambient pressure to complete the liquefaction process. The result is a mixture of gas and liquid that is sent to a separator. After the separator, the storing and discharging processes follow, which are the same for the four LAES technologies.

## Claude Process

The idea of the Claude process [18], which is shown in Fig. 9.3b, came from the Linde–Hampson process. The compression part of the Claude process is the same as the Linde–Hampson process. The difference occurs after the first heat exchanger, where the stream is split into two branches. One branch passes through the next two heat exchangers and is then expanded in a J–T valve to generate a gas and liquid mixture, which is subsequently phase separated. The other branch goes directly to an expander to produce refrigeration and electricity while expanding to ambient pressure. Then the expanded branch is mixed with the recirculating vapour to act as a cold stream to cool down the primary incoming air.

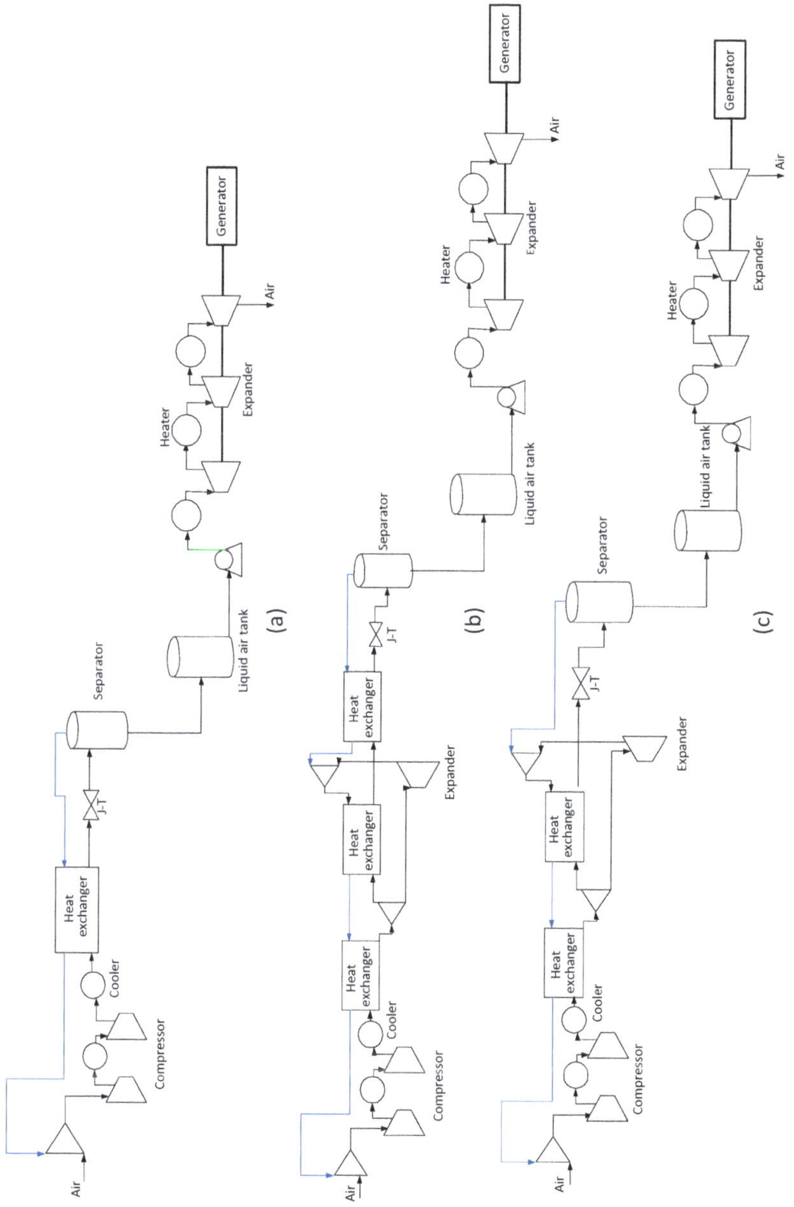


**Fig. 9.3** (**a**) Linde–Hampson process; (**b**) Claude process; (**c**) Kapitza process

## Kapitza Process

The Kapitza process is distinguished from the Claude process in that the third heat exchanger is removed [19], as shown in Fig. 9.3c. The recirculating vapour from the separator is mixed directly with the expanded branch.

## Modified Claude Process

The modified Claude process with hot and cold thermal energy storage has a different layout than the previous processes, as shown in Fig. 9.4. The ambient air is first compressed in a two-stage compressor to reach high pressure. The high-pressure air passes through two heat exchangers to obtain the energy level at low temperature from intermediate fluids, which are methanol and propane from the cold energy storage. Then the air enters a cryoturbine to expand to ambient pressure, which results in a gas and liquid mixture that is sent to a separator. For the process, both heat of compression and cold thermal energy (cooling) from regasification are stored in hot oil and methanol or propane, respectively. In this way, hot and cold thermal energy can be recycled to improve the efficiency of the overall system [20].

# Analysis and Discussion

In order to evaluate the performance of the different processes, a few parameters were selected:

- Liquid yield
- Round-trip efficiency
- Specific energy consumption
- Exergy efficiency

Liquid yield $\eta_{LA}$ is defined as the ratio of the mass flow rate of liquid air ($m_{liq}$) and the total mass flow rate of compressed air ($m_{comp}$) that includes the recycling of air:

$$\eta_{LA} = \frac{m_{liq}}{m_{comp}} \tag{9.2}$$

The most critical parameter is certainly the round-trip efficiency $\eta_{RT}$ that is defined as the work output ($W_{out}$) in recovery mode divided by the work input ($W_{in}$) in storage mode:

$$\eta_{RT} = \frac{W_{out}}{W_{in}} = \frac{m_{liq} W_T}{m_{comp} W_C} = \eta_{LA} \times \frac{W_T}{W_C} \tag{9.3}$$

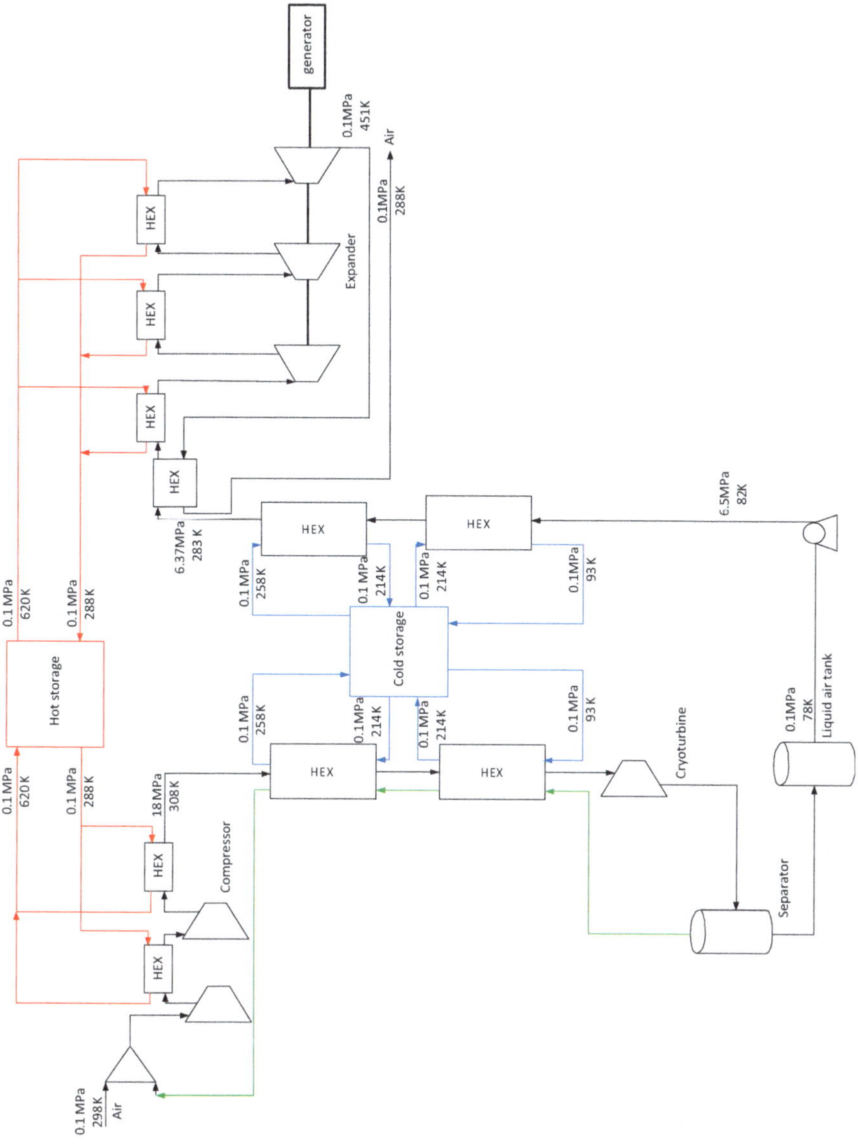


**Fig. 9.4** Modified Claude process with hot and cold thermal energy storage

where $w_T$ and $w_C$ represent the specific work (kJ/kg) of the expanders and the compressors, respectively. Another parameter for specific energy consumption (SEC) is the network consumed per mass of liquid air produced:

$$\text{SEC} = \frac{W_{\text{net}}}{m_{\text{liq}}} \tag{9.4}$$

where $W_{\text{net}}$ is calculated as follows:

$$W_{\text{net}} = \sum W_{\text{comp}} - \sum W_{\text{exp}} \tag{9.5}$$

where $W_{\text{comp}}$ and $W_{\text{exp}}$ are the work (kJ) of the compressors and expanders, respectively.

Finally, exergy efficiency could also be considered for comparing different processes. There are several approaches to calculate exergy efficiencies, such as input–output exergy efficiency [21], consumed–produced exergy efficiency [22] and exergy transfer effectiveness [23]. A reasonable method to evaluate exergy efficiency for sub-ambient processes, such as the liquid air energy storage system, is the exergy transfer effectiveness (ETE). The ETE is defined by using a classification of exergy sinks (produced exergy) and exergy sources (consumed exergy). As the name suggests, the ETE can reveal the exergy transfer within a process. In the LAES, there are three subprocesses: charging, storage and discharging, and it would be valuable to determine the exergy losses for the storage process (both the charging and storing process) and the discharging process. This information could be used to evaluate different sections of the system, and further directions for improvement could be pointed out based on the results of the analysis. The formula for exergy efficiency is given by the definition of the ETE:

$$\text{ETE} = \frac{\sum \text{Exergy sinks}}{\sum \text{Exergy sources}} \tag{9.6}$$

Marmolejo-Correa and Gundersen [23] proposed the ETE considering only thermo-mechanical exergy that was decomposed into temperature-based and pressure-based exergies, which quantify the exergy contributions from the temperature and pressure levels of the streams. The ETE even included shaft work as exergy sink or source depending on production or consumption. The thermo-mechanical exergy $E^{\text{TM}}$, temperature-based exergy $E^{\text{T}}$, and pressure-based exergy $E^{\text{P}}$ are calculated by Eqs. (9.7)–(9.10). Temperature-based exergy $E^{\text{T}}$ can be further classified into above ambient, across ambient and below ambient, since the exergy content of heat is estimated differently above and below ambient temperature, as shown by Eq. (9.11). Kim and Gundersen [24] further extended the ETE method to include chemical exergy related to chemical reactions and compositional changes. Therefore, the ETE is suitable for low-temperature processes, and in particular, provides accurate estimation for the case when the operating temperatures of the process range from below ambient to above ambient. The terms involved in the exergy sinks and sources of different LAES processes are listed in Table 9.2. It is assumed that the heat provided to the heaters in the three typical processes (Linde–Hampson, Claude and Kapitza) comes from a source with constant temperature of 360 °C.

$$E^{\mathrm{TM}} = H(T,p) - H(T_0,p_0) - T_0[S(T,p) - S(T_0,p_0)] \tag{9.7}$$

$$E^{\mathrm{TM}} = E^{\mathrm{T}} + E^{\mathrm{P}} \tag{9.8}$$

$$E^{\mathrm{T}} = H(T,p) - H(T_0,p) - T_0[S(T,p) - S(T_0,p)] \tag{9.9}$$

$$E^{\mathrm{P}} = H(T_0,p) - H(T_0,p_0) - T_0[S(T_0,p) - S(T_0,p_0)] \tag{9.10}$$

$$E[Q(T)] = Q\left[1 - \frac{T_0}{T}\right] \quad T \geq T_0 \tag{9.11}$$

$$E[Q(T)] = Q\left[\frac{T_0}{T} - 1\right] \quad T \leq T_0 \tag{9.12}$$

where $H$, $S$, $Q$, $T$ and $p$ are enthalpy (kJ/s), entropy (kJ/(kg K)), heat (kJ/s), temperature (K) and pressure (kPa) of the streams, respectively; $T_0$ and $p_0$ are ambient temperature and pressure (25 °C and 1 bar), respectively; $E(Q(T))$ is the exergy content of heat (kJ/kg). In Table 9.3, subscripts liq, hot, cold and heater represent the liquid air, hot oil, cold fluids (methanol and propane) and heaters in the processes.

Table 9.4 lists the liquid yield, specific energy consumption, exergy efficiency and round-trip efficiency for three typical liquefaction processes and the modified Claude process with hot and cold energy storage. The storage and discharging sections of the three first processes are the same; the only difference is the liquefaction section for each process. The comparison of the four processes in Table 9.4 shows that the modified Claude process has advantages in terms of liquid yield, specific energy consumption, exergy efficiency and round-trip efficiency. In fact, without considering investment cost, this process is superior compared to the Claude process and the Kapitza process. This is because the modified Claude process has a relatively small recirculation ratio, which is partly related to the use of hot and cold

**Table 9.3** The terms for calculating ETE of different LAES processes

| | Charging and storage process | | Discharging process | |
|---|---|---|---|---|
| LAES processes | Exergy sinks | Exergy sources | Exergy sinks | Exergy sources |
| Linde–Hampson | $E_{\mathrm{liq}}$ | $w_c$ | $w_t$ | $E_{\mathrm{liq}} + E(Q(T))_{\mathrm{heater}}$ |
| Claude | $E_{\mathrm{liq}}$ | $w_c$ | $w_t$ | $E_{\mathrm{liq}} + E(Q(T))_{\mathrm{heater}}$ |
| Kapitza | $E_{\mathrm{liq}}$ | $w_c$ | $w_t$ | $E_{\mathrm{liq}} + E(Q(T))_{\mathrm{heater}}$ |
| Modified Claude | $E_{\mathrm{liq}} + E(Q(T))_{\mathrm{hot}}$ | $w_c + E(Q(T))_{\mathrm{cold}}$ | $w_t + E(Q(T))_{\mathrm{cold}}$ | $E_{\mathrm{liq}} + E(Q(T))_{\mathrm{hot}}$ |

**Table 9.4** Performance parameters for liquid air energy storage processes

| Performance | | Linde–Hampson | Claude | Kapitza | Modified Claude |
|---|---|---|---|---|---|
| Liquid yield (%) | | 7.32 | 16.58 | 16.13 | 85.34 |
| Specific energy consumption (kWh/kg) | | 3.83 | 0.81 | 0.81 | 0.12 |
| Exergy efficiency (%) | Charging and storage | 5.1 | 21.12 | 21.16 | 84.12 |
| | Discharging | 38.64 | 38.64 | 38.64 | 78.31 |
| Round-trip efficiency (%) | | 4.29 | 15.01 | 15.04 | 55.43 |

thermal energy storages in the process. The Linde–Hampson process has the lowest score for all performance indicators except exergy efficiency for discharging.

## Future Prospects

Thanks to its unique features, liquid air energy storage (LAES) overcomes the drawbacks of pumped hydroelectric energy storage (PHES) and compressed air energy storage (CAES). It is not geographically constrained; it uses commercially available equipment (thus reduced upfront costs), and it integrates well with traditional power plants. Therefore, a LAES system can probably be considered as a viable option for grid-scale (hundreds of MW) electric energy storage, even in a stand-alone configuration. Until now, the most significant application scale for LAES is a small pilot plant with 5 MW storage capacity, which was built by Highview Power in the United Kingdom [25]. The round-trip efficiency of the process was around 0.6. Li et al. [26] integrated the LAES with a nuclear power plant (NPP), and the round-trip efficiency was claimed to reach 0.7. An approach with a liquid air Rankine cycle was proposed by Ameel et al. [27], and the round-trip efficiency of 0.43 could be achieved. A stand-alone LAES plant was simulated by Guizzi et al. [20] reaching the round-trip efficiency of 0.55. None of the present studies discussed has been applied in large scale. Thus, LAES first needs further research to increase overall efficiency, store hot and cold thermal energy efficiently, and increase response time. Then the implementation of the technology in practical applications could be considered.

The future applications of LAES systems could be to integrate renewable energy, waste heat and batteries to form a hybrid system [28]. The hybrid plant concept integrating high speed drives for compressor and expander connections with small electrochemical storage makes these systems a promising solution, even when fast dynamic responses are required. This feature, together with a high round-trip efficiency, are becoming essential performance requirements to manage a power system where the amount of renewable energy sources is rapidly growing, and where the number of power generating units that can offer control services is decreasing. Therefore, hybrid power plants based on LAES technology may be a promising solution to store energy and use it at peak times with satisfactory performance.

**Acknowledgements** The authors are grateful to the ENERSENSE programme and NTNU Team Hydrogen at the Norwegian University of Science and Technology (NTNU) for supporting and helping on this book project.

## References

1. Blyth W, Lefevre N (2004) Energy security and climate change policy interactions: an assessment framework. International Energy Agency (IEA), Paris, France
2. Burheim OS (2017) Engineering energy storage. Academic, London
3. Kara M, Hirvonen R, Mattila L, Viinikainen S, Tuhkanen S (2001) Energy visions 2030 for Finland. Edita, Helsink
4. Alanne K, Saari A (2006) Distributed energy generation and sustainable development. Renew Sustain Energy Rev 10(6):539–558
5. Arteconi A, Hewitt NJ, Polonara F (2012) State of the art of thermal storage for demand-side management. Appl Energy 93:371–389
6. Aneke M, Wang M (2016) Energy storage technologies and real life applications—a state of the art review. Appl Energy 179:350–377
7. Dunn B, Kamath H, Tarascon J-M (2011) Electrical energy storage for the grid: a battery of choices. Science 334(6058):928–935
8. Fernandez A, Martínez M, Segarra M, Martorell I, Cabeza L (2010) Selection of materials with potential in sensible thermal energy storage. Sol Energy Mater Sol Cells 94(10):1723–1729
9. Chen H, Cong TN, Yang W, Tan C, Li Y, Ding Y (2009) Progress in electrical energy storage system: a critical review. Prog Nat Sci 19(3):291–312
10. Pardo P, Deydier A, Anxionnaz-Minvielle Z, Rougé S, Cabassud M, Cognet P (2014) A review on high temperature thermochemical heat energy storage. Renew Sustain Energy Rev 32:591–610
11. Rosen MA (1992) Appropriate thermodynamic performance measures for closed systems for thermal energy storage. J Sol Energy Eng 114(2):100–105
12. Liew PY, Theo WL, Wan Alwi SR, Lim JS, Abdul Manan Z, Klemeš JJ et al (2017) Total Site Heat Integration planning and design for industrial, urban and renewable systems. Renew Sustain Energy Rev 68:964–985
13. Ibrahim H, Ilinca A, Perron J (2008) Energy storage systems—characteristics and comparisons. Renew Sustain Energy Rev 12(5):1221–1250
14. Bullough C, Gatzen C, Jakiel C, Koller M, Nowi A, Zunft S (2004) Advanced adiabatic compressed air energy storage for the integration of wind energy. In: Conference advanced adiabatic compressed air energy storage for the integration of wind energy, London, UK
15. Mahlia TMI, Saktisahdan TJ, Jannifar A, Hasan MH, Matseelar HSC (2014) A review of available methods and development on energy storage; technology update. Renew Sustain Energy Rev 33:532–545
16. Biris AS, Biris AR, Lupu D, Buzatu D, Darsey J, Mazumder MK (2004) Use of carbon nanostructures for hydrogen storage for environmentally safe automotive applications. In: Conference record of the 2004 IEEE industry applications conference. 39th IAS annual meeting, vol 2, pp 953–956
17. Linde C (1903) Process of producing low temperatures, the liquefaction of gases, and the separation of the constituents of gaseous mixtures. US Patents 727650
18. Claude G (1913) Process for separation of the constituents of gaseous mixtures. US Patents 607775
19. Kapitza PL (1946) Method and means for distillation of low boiling point liquids. US Patents 625107
20. Guizzi GL, Manno M, Tolomei LM, Vitali RM (2015) Thermodynamic analysis of a liquid air energy storage system. Energy 93:1639–1647

21. Lior N, Zhang N (2007) Energy, exergy, and second law performance criteria. Energy 32 (4):281–296
22. Kotas TJ (2012) The exergy method of thermal plant analysis: Exergon/Paragon, London
23. Marmolejo-Correa D, Gundersen T (2015) A new efficiency parameter for exergy analysis in low temperature processes. Int J Exergy 17(2):135–170
24. Kim D, Gundersen T (2018) Development and use of exergy efficiency for complex cryogenic processes. Energy Convers Manag 171:890–902
25. Highview Power Storage official site. http://www.highview-power.com. Last accessed 04.06.19
26. Li Y, Cao H, Wang S, Jin Y, Li D, Wang X et al (2014) Load shifting of nuclear power plants using cryogenic energy storage technology. Appl Energy 113:1710–1716
27. Ameel B, T'Joen C, De Kerpel K, De Jaeger P, Huisseune H, Van Belleghem M et al (2013) Thermodynamic analysis of energy storage with a liquid air Rankine cycle. Appl Therm Eng 52 (1):130–140
28. Antonelli M, Barsali S, Desideri U, Giglioli R, Paganucci F, Pasini G (2017) Liquid air energy storage: potential and challenges of hybrid power plants. Appl Energy 194:522–529

# Hydrogen and Biogas 10

Eline F. J. Gregorie, Jacob J. Lamb, Kristian M. Lien, Bruno G. Pollet, and Odne S. Burheim

## Introduction

Among the renewable energy sources that are essential to face challenges such as climate change and energy depletion, biogas has become one of the attractive pathways in recent years [1]. Biogas derives from the natural degradation of organic materials by microorganism anaerobic digestion [2–5]. These organic materials come from waste feedstocks considered as renewable sources since the rejection of waste is continual. On top of being a renewable substitute for fossil fuels, biogas helps in waste management. In this way, agricultural and industrial residues, municipal organic wastes, and sewage sludge are common sources of feedstock including seeds, grains, and sugars, lignocellulosic biomass like crop residues and woody crops, or algae with their high carbohydrate content [6].

E. F. J. Gregorie
Department of Energy and Process Engineering, Norwegian University of Science and Technology, Trondheim, Norway

J. J. Lamb (✉)
Department of Electronic Systems, ENERSENSE, Norwegian University of Science and Technology, Trondheim, Norway

Department of Energy and Process Engineering, ENERSENSE, Norwegian University of Science and Technology, Trondheim, Norway
e-mail: jacob.j.lamb@ntnu.no

K. M. Lien
Department of Energy and Process Engineering, ENERSENSE, Norwegian University of Science and Technology, Trondheim, Norway
e-mail: Kristian.m.lien@ntnu.no

B. G. Pollet · O. S. Burheim
Department of Energy and Process Engineering, ENERSENSE and NTNU Team Hydrogen, Norwegian University of Science and Technology, Trondheim, Norway
e-mail: bruno.g.pollet@ntnu.no; odne.s.burheim@ntnu.no



J. J. Lamb, B. G. Pollet (eds.), *Micro-Optics and Energy*,
https://doi.org/10.1007/978-3-030-43676-6_10

The traditional biological process for anaerobic digestion is composed of four stages:

1. Hydrolysis: where the disintegration of the feedstock into simple sugars, amino acids, and fatty acids occurs through the action of hydrolytic bacteria
2. Acidogenesis: where the previous products are transformed into short-chain organic acids and alcohols producing $H_2$ and $CO_2$
3. Acetogenesis: where $H_2$ and $CO_2$ are reduced to acetic acid by the action of homoacetogenic microorganisms
4. Methanogenesis: where the production of methane occurs by methanogenic bacteria under anaerobic conditions [7]

The raw biogas produced is mainly composed of methane $CH_4$ (between 55 and 70%) and $CO_2$ (between 30 and 45%), in association with trace components such as $H_2S$, ammonia, $H_2$, nitrogen, $O_2$, vapor water, siloxanes, CO and hydrocarbons [8, 9]. The presence of these components and their respective quantities depends largely on the nature of the biogas sources and the operational conditions used during the process.

In order to be used for many applications, biogas has to then meet cleaning processes to remove the trace components and upgrading processes to increase the fuel quality. The presence of trace components might be inconvenient for further uses, and many applications require control of the level of impurities. Many technologies have been developed for biogas purification, and the selection of the appropriate technology depends on the utilisation requirements. Condensation methods such as demisters, cyclone separators or moisture traps and drying methods such as adsorption or absorption can be applied to remove water. To remove $H_2S$, some techniques are used during anaerobic digestion like air dosing to the biogas and addition of iron chloride into the digester tank, while some other techniques can be used after the digestion process like adsorption and absorption. During these treatments, other impurities like ammonia, nitrogen, $O_2$, siloxanes, CO and hydrocarbons can be removed; however, if the level of these impurities is still too high, some additional treatments might be applied [10].

Then the separation of $CH_4$ from $CO_2$ must be operated to adjust the calorific value of the biogas in order to meet the standards for use since $CH_4$ is the component that contributes to the heating value giving the energy content. For example, 1 $m^3$ of raw biogas with a $CH_4$ content of 60% has a lower heating value of 21.5 MJ compared to the one of 35.8 MJ corresponding to 1 $m^3$ of pure $CH_4$ at standard temperature and pressure [11]. Separation methods include water scrubbing, cryogenic separation, physical or chemical $CO_2$ absorption, pressure swing adsorption, membrane separation or biological methods [9].

The utilisation of biogas includes traditional pathways, which use methane present in biogas as combustible component to produce heat, power, or to be used as transport fuel. In Europe about 50% of total biogas consumption was used for heat production in 2015 [1]. This includes mostly cooking and heating applications, with the direct use of biomethane as fuel for stoves and boilers. Boilers have high tolerance to biogas quality. However, to replace natural gas in domestic stoves, the $CH_4$ content of the biogas should be close to 100%, and only low levels of $H_2S$ are tolerated since $H_2S$ has toxicity and corrosion issues [9].

The worldwide biopower production in 2015 represented installed capacity of about 106 GW, among the 1969 GW of installed power capacity of renewable energy [1]. Biogas as fuel for engines and gas turbines for producing electricity has the benefit to be able to be used without upgrading step since internal engines can be supplied with a mixture of $CH_4$ and $CO_2$; however, the level of $H_2S$ in the feed biogas should also be controlled for this kind of use.

Combined heat and power plants, with their high efficiency under optimal conditions of 90% including 40% for electricity generation and 50% for thermal production, also accommodate biogas as a promising and economical fuel [6].

As a substitute for natural gas, biogas can be injected into natural gas networks as soon as it achieves the standards of natural gas. These require high $CH_4$ proportions (93–95% and 70–98% according to US and EU requirements, respectively) and a maximum $H_2S$ content to avoid equipment erosion (6–88 mg/m$^3$ and 2–15 mg/m$^3$ for the US and EU grids, respectively) [6].

Upgraded biogas as a vehicle fuel is another traditional pathway, achieving about 4% of global road transport fuel in 2015 [1]. The purified biogas is compressed to become compressed biogas (CBG) that can be directly used in natural gas vehicles. Moreover, it has similar properties to compressed natural gas (CNG). Rigid containers are usually used for storage and distribution at a pressure between 200 and 250 bar [12]. It can also be liquefied cryogenically into liquefied biomethane (LBM), used as a substitute for liquefied natural gas (LNG) and be delivered using tankers [13].

Current utilisations have been explored in recent years, where biogas can be used as starting material to form higher-value products. It is the case for synthesis gas production from biogas reforming, which forms a mixture of CO and $H_2$ and enables further chemical productions such as methanol used in pharmaceutical and textile industries, or other higher hydrocarbons and alcohols [14].

Due to its flexibility in use and its ability to be stored, biogas has an important role to play for balancing the energy grids in the future. In this chapter, we will go deeper on the renewable energy applications coming from the integration pathways between biogas and $H_2$. $H_2$ also has promising issues and will enable further renewable energy applications from biogas, whose relevance will be driven by the relative price between biogas and electricity. Therefore, this chapter will describe the potential of the biogas reforming and its renewable $H_2$ production issue, relevant in the case of high demand for energy and high electricity/biogas relative price. Then we will present the energy storage possibilities that the meeting of $H_2$ and biogas technologies enable, pertinent to store the cheap electricity surplus coming from the integration of wind and solar renewable energy into the energy grid.

## Biogas Reforming for Hydrogen Production

There are many processes for $H_2$ production. We can separate them into two distinct categories: the conventional ways that produces $H_2$ from fossil fuel such as natural gas or coal through hydrocarbon reforming and hydrocarbon pyrolysis, and the alternative ways that use renewable sources. The alternative paths produce $H_2$ using either biomass processes with biomass as raw material, that can be biological (bio-photolysis, dark fermentation and photo fermentation) or thermochemical processes (pyrolysis, gasification, combustion and liquefaction), or water splitting with water as raw material that includes electrolysis, thermolysis and photolysis technologies [15]. These alternative methods are not yet fully competitive with conventional methods that dominate the large-scale $H_2$ production industries, with the dominance of natural gas reforming [16].

Despite this, the dominant conventional method promotes the reforming of light hydrocarbons, such as methane. This is interesting since methane is a significant biogas component and biogas is considered as a renewable source from waste. Therefore, the production of $H_2$ using biogas reforming appears as an interesting issue to present here for $H_2$ production with the integration of a renewable energy source using conventional methods.

Methane from biogas can be an energy source to produce $H_2$ through reforming techniques. These processes convert the methane into a mix of CO and $H_2$, called synthesis gas (syngas). Syngas is a high-value product that can be used directly in chemical engineering, but its separation as a step of pure $H_2$ production is an issue.

Some of the reforming techniques are conventional and already well-known by industries for large-scale $H_2$ production using natural gas as hydrocarbon sources such as steam reforming (SR), partial oxidation reforming (POR) and autothermal reforming (ATR) [15]. These reaction procedures use $H_2O$ and $O_2$ as sources of $O_2$. Some other reforming processes consist of producing $H_2$ and CO using $CO_2$ as source of $O_2$, such as dry reforming (DR). These techniques are particularly attractive regarding biogas reforming since raw biogas is composed of both $CH_4$ and $CO_2$ reactants.

Whatever the reforming process, extra steps are then necessary to obtain pure $H_2$. The CO removal is operated through a shift reaction commonly done in a conversion reactor, converting the CO into $CO_2$ and an extra amount of $H_2$ [8]. The final $H_2$ purification is then usually operated through pressure swing adsorption (PSA) in a separation unit [15].

## Reforming Techniques for Hydrogen Production

Reforming processes can be operated in either tubular fixed-bed or fluidised bed reactors [17]. Continuous fixed-bed reactors are widely used since they have relatively low costs and simple architectures [6]. A different kind of reforming can be operated in these reactors, which is summarised in Table 10.1. In general, all of these

**Table 10.1** Overview of reforming processes [8, 15, 17, 18]

| Process | Equation | $\Delta H_{298}$ (kJ/mol) | Temperature range (°C) | $H_2/CO$ ratio | Major features |
|---|---|---|---|---|---|
| SR | $CH_4 + H_2O \leftrightarrow CO + 3\ H_2$ | 206 | 840–950 | 3 | External heat source required |
| POR | $CH_4 + \frac{1}{2}O_2 \leftrightarrow CO + 2\ H_2$ | –35.6 | 700–900 | 2 | No external heat source required, complete combustion to avoid |
| ATR | $CH_4 + \frac{1}{2}H_2O + \frac{1}{4}O_2 \leftrightarrow CO + \frac{5}{2}H_2$ | 0 | 500–700 | 2.5 | No external heat source required |
| DR | $CH_4 + CO_2 \leftrightarrow 2CO + 2H_2$ | 247 | 640–900 | 1 | Consumes both $CH_4$ and $CO_2$, external heat source required, high coke accumulation |
| DOR | $CH_4 + \frac{1}{2}CO_2 + \frac{1}{4}O_2 \leftrightarrow \frac{3}{2}CO + 2H_2$ | 123.5 | 750–900 | 1.3 | Reduced external heat source required |

techniques can be conducted at low pressure, but in practice, intermediate pressures between 10 and 20 bars are used to improve the efficiency [8, 18].

## Steam Reforming Process (SR)

For the last century, $H_2$ production on an industrial scale mainly comes from methane steam reforming [17, 19].

$$CH_4 + H_2O \rightarrow CO + 3H_2 \quad \Delta H_{298} = 206\ \mathrm{kJ/mol} \tag{10.1}$$

It is the catalytic conversion of the methane molecule, which is dissociated with the presence of water vapour and heat, to produce $H_2$ and CO. The $H_2$/CO ratio is equal to three, meaning an $H_2$ yield of approximately 75%. Methane needs a high temperature (840–950 °C) in order to react. This temperature can be decreased using an appropriate catalyst [13]. The tremendous heat consumption required to shift the reaction equilibrium towards the right to yield $H_2$ and CO is the main disadvantage of this process since the reaction is highly endothermic ($\Delta H_{298} >> 0$) and an external heat source is required. Usually, external combustion in a power unit fed with the biogas is operated in order to supply heat [20].

The most common process uses Ni-based catalysts on ceramic supports [19]. Nickel has excellent catalytic properties and is an affordable and available resource. As a result, many Ni-based catalysts have been developed in order to improve the system performance in term of reactivity, selectivity and lifetime [18].

The operational conditions of steam reforming lead to parallel carbon formation reactions, such as methane decomposition reaction (Eq. 10.2), Boudouard reaction (Eq. 10.3) or disproportionation reduction of CO (Eq. 10.4) [8].

$$CH_4 \rightarrow C + 2H_2 \tag{10.2}$$

$$2CO \rightarrow C + CO_2 \tag{10.3}$$

$$CO + H_2 \rightarrow C + H_2O \tag{10.4}$$

These carbon formations are a problem since coke (a final carbon-rich solid material; represented by "C" in Eqs. 10.2, 10.3 and 10.4) deposits on the Ni-catalyst surface leading to its deactivation. Basic systems can limit the coke deposition since they promote carbon gasification in water improving its adsorption properties [18]. Combinations of elements to form catalysts with oxides that have basic properties (e.g. magnesium, calcium and potassium) have been investigated. Some other catalysts, like Pt- and Pd-based catalysts, are more stable regarding coke formation [8]. Ruthenium is also widely used in methane SR due to its high activity, high $H_2$ selectivity and excellent stability [18]. Another way to reduce the coke deposition due to the reaction (Eq. 10.4) is to operate under high partial pressure of $H_2O$ in order to shift the equilibrium towards the right and limit carbon formation [21].

## Partial Oxidation Reforming Process (POR)

Partial oxidation (Eq. 10.5) is another way to produce $H_2$ via methane reforming [7].

$$CH_4 + \frac{1}{2}O_2 \rightarrow CO + 2H_2 \quad \Delta H_{298} = -35.6\ \mathrm{kJ/mol} \tag{10.5}$$

A methane molecule is partially oxidised to produce $H_2$ and CO. The $H_2/CO$ ratio is equal to two, meaning an $H_2$ yield of approximately 67%. The process requires a temperature range from 700 to 900 °C [7]. The molar $O_2/CH_4$ ratio is a crucial feature to control for this process, since an $O_2/CH_4$ ratio higher than 0.5 would lead to complete combustion of the methane [8].

$$CH_4 + 2O_2 \rightarrow CO_2 + 2H_2O \quad \Delta H_{298} = -801.7\ \mathrm{kJ/mol} \tag{10.6}$$

This oxidation reaction is highly exothermic and leads to a temperature increase. Therefore, to reduce the amount of $H_2$ produced, complete combustion reactions would lead to hotspots in the reactor bed that facilitate coke formation on the catalyst surface [19]. Ni-based catalysts are commonly used, with importance on the selectivity to favour the partial oxidation reaction. Some highly active and selective catalysts have been reported for POR of methane, such as NiO-MgO, Ni-Mg-Cr-La-O and Ca-Sr-Ti-Ni solid solutions [8]. Among the best catalysts for this reaction, Ru-based catalysts have a high coke resistance at high temperature. Ruthenium oxide improves the methane conversion rate and the selectivity for $H_2$ production since it has an excellent thermal and chemical stability and high resistance to chemical corrosion [18].

## Autothermal Reforming Process (ATR)

In order to reduce the drawback of the endothermic nature of the SR, and the disadvantage of the lower $H_2$ yield of the POR, autothermal reforming (Eq. 10.7) as a combination of both processes has been developed [19].

$$CH_4 + \frac{1}{2}H_2O + \frac{1}{4}O_2 \rightarrow CO + \frac{5}{2}H_2 \quad \Delta H_{298} = -0\ \mathrm{kJ/mol} \tag{10.7}$$

Steam and $O_2$ are injected into the reformer to cause both reforming and oxidation reactions, simultaneously. The $H_2/CO$ ratio can be adjusted with the inlet ratios, and the maximum $H_2$ yield is about 74% (corresponding to an $H_2/CO$ ratio of 2.8) [15]. The term "autothermal" refers here to the fact that the process does not require any external heat supply. Autothermal reforming balances the heat demand of the endothermic reactions with the heat released from the exothermic reactions and can be considered as an efficient system [19].

The partial oxidation takes place in a thermal zone in the reactor, generating heat for the steam reforming that is conducted in a catalytic zone [8]. The highest plant efficiencies for biogas reforming via ATR are near 75%, for a temperature range from 500 to 700 °C and an O/C ratio from 0.8 to 0.9 [22]. A maximum temperature of 700 °C is usually applied since a lower temperature increases the $H_2/CO$ ratio due to the increased content of the non-reacted methane. Additionally, a lower temperature is more suitable for further CO removal [23].

## Dry Reforming Process (DR)

In this process, the $H_2$ production derives from the reaction of methane with $CO_2$ [18].

$$CH_4 + CO_2 \rightarrow 2CO + 2H_2 \quad \Delta H_{298} = 247\ \mathrm{kJ/mol} \tag{10.8}$$

This reaction occurs at a temperature higher than 640 °C [18]. This high endothermic reaction leads to the need for an external heat source. This process has many attractive aspects since $CH_4$ and $CO_2$ are the two primary compounds of raw biogas.

Dry reforming leads to an $H_2/CO$ ratio close to 1, meaning $H_2$ yields of around 50%. The $H_2/CO$ ratio makes this reaction particularly interesting regarding requirements for further synthesis gas utilisation like the Fischer–Tropsch synthesis [8]. Some parallel reactions could also occur in this process that would modify the equilibrium conversion. It is the case for the reverse gas-water shift (Eq. 10.9) that can lead to a decrease in the $H_2$ yield.

$$CH_2 + H_2 \rightarrow CO + H_2O \quad \Delta H_{298} = 41.2\ \mathrm{kJ/mol} \tag{10.9}$$

Other reactions leading to coke formation previously seen like the Boudouard reaction (Eq. 10.3) and the methane decomposition (Eq. 10.2) could also modify the conversion rate [8]. The lower H/C ratio in DR compared to previous reforming processes leads to more coke formation [18]. Coke accumulation on the support and the active phase of the catalyst is favoured in DR due to the carbon–water steam reaction (Eq. 10.10), as the reverse disproportionation reduction of CO (Eq. 10.4) cannot occur as it does with SR [8].

$$H_2O + C \rightarrow CO + H_2 \tag{10.10}$$

The major problem of DR is the coke accumulation that causes catalyst deactivation and could block the reactor. As for previous processes, catalyst materials with $H_2$ selectivity and thermal stability are used with great interest in the ones that can reduce carbon deposition [8]. Ru-, Rh- and Pt-based catalysts increase the $H_2$ production performance, but due to their high costs and limited availability they are not developed at industrial level, whereas Ni- and Co-based catalysts are commonly used [8, 18].

A circulating bed reactor with a regenerator is a technical way often used to deal with the coke deposition on the catalyst, which allows the process to avoid catalyst deactivation through a continuous operation [20]. The coke accumulated on the catalyst is burnt and eliminated, and the heat released can be used by the endothermic dry reforming.

### Dry Oxidation Reforming (DOR)

In order to control the carbon deposition on the surface of the catalyst, dry oxidation reforming (Eq. 10.11) as the combination of dry reforming and partial oxidation reforming has been developed [8].

$$CH_4 + \frac{1}{2}CO_2 + \frac{1}{4}O_2 \rightarrow \frac{3}{2}CO + 2H_2 \quad \Delta H_{298} = 123.5\ \mathrm{kJ/mol} \qquad (10.11)$$

This process also reduces the energy demand for the reaction, since the exothermic partial oxidation releases heat useful for the endothermic dry oxidation. The $H_2$/CO ratio also increases leading to an $H_2$ yield close to 60%.

## Hydrogen Purification Processes

Regarding $H_2$ utilisation, the gas purification is often required. For example, a low CO content is necessary in order to avoid poisoning of the platinum-alloy catalyst in the transportation sector with PEMFCs [25].

### Condenser Unit

In order to remove the water from the produced syngas, the output stream from the reactor can be cooled down through a condenser [20].

### Water-Gas Shift Reaction (WGS)

The water-gas-shift (WGS) reaction (Eq. 10.12) is the standard way to eliminate CO from the produced syngas [8].

$$CO + H_2O \rightarrow CO_2 + H_2 \quad \Delta H_{298} = -41.2\ \mathrm{kJ/mol} \qquad (10.12)$$

This reaction enables extra $H_2$ production. It operates at a temperature range from 140 to 350 °C [24], since very high temperatures above 800 °C favour the reverse water-gas shift and would decrease the $H_2$ yield [19].

Catalysts commonly used for the water-gas shift reaction are Fe, Cu, Mo or Fe-Pd alloys [8]. This process can be decomposed into two steps to increase the $H_2$ content of the syngas. This is done through a high-temperature step (HTS) and a low-temperature step (LTS). An $H_2O/CO$ ratio of 3 is commonly used, and this process involves operating temperatures of around 350 °C for the HTS conversion reactor with a Fe-Cr catalyst, whereas the LTS one commonly works around 220 °C with a Cu-Zn catalyst [20]. In the case of steam reforming, this step can be directly conducted in the reformer [18], leading to a combination of both steam reforming and water-gas shift reactions [8].

$$CH_4 + 2H_2O \rightarrow CO_2 + 4H_2 \quad \Delta H_{298} = 164.4\ \mathrm{kJ/mol} \tag{10.13}$$

## Pressure Swing Adsorption (PSA)

To archive the target purity above 99.99% of $H_2$, the WGS reaction is followed by a pressure swing adsorption (PSA) [20]. This method is based on differences in Van Der Waals forces between $CO_2$ and $H_2$ gases and the surface of adsorbents and is conduct in an adsorption column [8]. The PSA unit works under elevated pressure, from around 20 bar to 1.5 bar of the outflow. The compression of the gas injected, which requires a large amount of energy, is a necessary step if it is not done previously [23]. If some trace compounds are present, such as $H_2O$ or unreacted $CH_4$ and CO, this step also enables their removal.

The previous utilisation of an LT-WGS step, which enables an increase of the $H_2$ concentration, leads to advantages for the PSA operating conditions and the overall system efficiency [17]. When the adsorbents are saturated with $CO_2$, the gas is treated in another column, and the pressure in the saturated column is reduced in order to enable $CO_2$ desorption. Therefore, to obtain continuous PSA operation it is necessary to have several PSA columns that can be opened and closed in sequence [6].

Conventional adsorbents are generally made from activated carbon. Some recent adsorbents include the application of Metal-Organic Framework (MOF) that can absorb/desorb $CH_4$, CO and $CO_2$ and could enable the direct methane reforming off-gas purification.

## Membrane Reactors

At low pressures, an undesirable reaction that could occur is the reverse water-gas shift (Eq. 10.9). This would transform the desired $H_2$ into CO and water. To avoid it, the removal of $H_2$ from the reformer reactor as soon as it is produced appears as a solution [25]. In this way, the introduction of a membrane into the reactor is relevant. This membrane allows a pure $H_2$ output flow. Packed bed membrane reactors and fluidised bed membrane reactors are alternative ways that have shown to work well with Pd-based thin membranes [26]. This kind of palladium membrane is at early phase of industrialisation. An example of membranes used for dry reforming in a fluidised bed is the permselective Pd/Ag membrane [25].

The utilisation of a membrane leads to increases in the productivity of $H_2$ since the removal from the reaction medium shifts the chemical equilibrium to form more products. Therefore, the reforming process can operate at lower temperature below 500 °C without any losses in methane conversion [8]. Compared to traditional reactors, the use of a membrane permeable to $H_2$ also enables operation at lower pressures with higher methane conversion rates in a reduced total reactor volume [27]. An inconvenience of this technique is that it increases the catalyst deactivation since the $H_2$ removal from the reactor also leads to an increase of coke accumulation [25]. Two-zone fluidised bed reactors can avoid this disadvantage, allowing the catalyst regeneration in a single unit [25].

## Biogas as Source for Reforming: The Influence of Impurities in Biogas

### Hydrogen Sulphide

Sulphur in biogas is mainly present in the form of $H_2S$ [9]. The presence of $H_2S$ impacts the catalyst activity and stability, even at minor concentrations. The sulphur from $H_2S$ reacts with active sites of the catalyst and limits the access to the reactants, reducing the catalyst activity.

Moreover, this restraint of the highly endothermic reactions like SR and DR increases the risk of catalyst deactivation resulting in overheating. In the case of Ni-based catalysts, the $H_2S$ acts as a poison and raises catalyst deactivation [19].

Furthermore, sulphur causes equipment corrosion [6]. Some sulphur-resistant catalysts are studied, but the previous desulfurisation of the biogas leading to a low ppm $H_2S$ content is required to improve the efficiency of the methane reformer and the lifetime of the catalysts. A solution is to implement a carbon bed at the entrance of the process and to use a cascade of beds with a first bed named guard bed as a sacrificial component to reduce the cost of upstream desulfurisation and separation [28].

### Oxygen

It is expected that the $O_2$ present in the fed biogas takes part in the POR, but as seen before a too high $O_2/CH_4$ ratio would lead to methane combustion, decreasing the $H_2$ yield. Some studies on dry reforming report that the presence of $O_2$ improves catalytic activity and stability. This improvement is because $O_2$ changes the reducing environment in which small Ni particles were unstable towards sintering. $O_2$ can inhibit sintering [6], and decreases carbon deposition leading to benefits regarding the catalyst activity and its lifetime [6]; however, the content of less than 5% of $O_2$ is recommended for efficient dry biogas reforming, since concentrations higher than 10% would lead to methane combustion and a decrease in $H_2$ production [6].

## Siloxanes

Siloxanes present as impurities in biogas can decompose and form $SiO_2$ with $O_2$ at high temperatures. These products accumulate on surfaces damaging the equipment. Siloxanes can also decrease the efficiency of the reforming and act as poison for steam reforming catalysts [6]. Therefore, siloxane removal before biogas reforming is required.

## Example of Plant and Economic Analysis

A conventional bio-$H_2$ energy system is composed of four steps: a reformer where the methane reforming process occurs; a conversion reactor for the water-gas shift reaction to operate with the two high and low temperature steps; a separation unit that enables the pressure swing adsorption; and a power generation unit that can be fed with the biogas as combustible [15].

A circulating bed reactor with a regenerator is commonly used to avoid catalyst deactivation due to the coke deposition. The efficiency of the system can be improved thanks to its catalyst regenerator since the coke is burnt, and the endothermic reforming reaction can use the released heat. Moreover, the hot off-gas from the regenerator can be used to produce steam through the heat recovery steam generator of the power generation unit, useful for the WGS process [20]. In order to improve the efficiency of the process in the case of dry reforming, the PSA off-gas composed of $CO_2$ can be fed to the reformer [20].

Studies on such a system show an exegetic efficiency of 55%, with the dry reforming process and the power generation as significant sources of irreversibility. The recirculation of the $CO_2$ PSA off-gas improves the system's exegetic performance, which is not possible in the case of steam reforming [20].

Steam reforming using biogas for $H_2$ production is reported to have similar levels of efficiency compared to the use of natural gas. An economic analysis in 2013 showed that US$0.27/kWh is achieved with a payback period of 8 years, considering investment, operation and maintenance costs [24]. Considering the lower heating value of $H_2$ (about 120 MJ/kg), this corresponds to a result of US$8.98/$kg_{H2}$ ($\approx$ 8 €/$kg_{H2}$) [24].

More recent results for steam reforming involving membrane reactors show that the adoption of membrane reactors increases the system efficiency by more than 20%. Indeed, system efficiencies are reported between 25 and 28% with an S/C ratio of 3 for a conventional ATR process and between 46 and 52% with S/C ratio of 4 for a conventional SR, whereas the membrane reactors show efficiencies around 70% with S/C ratio of 3 [17].

Regarding the economic analysis, $H_2$ production with a membrane reactor costs less than conventional reforming-WGS-PSA processes. In 2018, it was reported that the $H_2$ costs are 4.0 €/$kg_{H2}$ for a fluidised membrane reactor process producing

100 $kg_{H2}$/day at 20 bars, whereas 4.2 €/$kg_{H2}$ and 6.4 €/$kg_{H2}$ are respectively reported for the SR and ATR conventional processes at the same conditions [17]. It is essential to keep in mind that the $H_2$ production costs depend on the biogas feedstock since the methane content of the biogas depends on it.

Since biogas has higher costs of generation than natural gas, respectively 0.07 €/kWh compared to 0.03 €/kWh [29], the relevance from an economic point of view has to take into consideration together with the scale of the technology and the carbon price. Conventional techniques of reforming have been developed for natural gas reforming at large scale since the large global natural gas production, which reaches nearly 3750 billion $m^3$ in 2017 [30], is located at specific sites (leading to centralised $H_2$ production).

Decentralised $H_2$ production would minimise the need for distribution infrastructure and reforming techniques such as dry reforming, and membrane reactors are more competitive or feasible at a lower production scales. Since biogas has a decentralised feature compared to natural gas and a smaller global production (nearly 36 billion $m^3$ of methane in 2015 [1]), it appears more appropriate for decentralised $H_2$ production. Additionally, decentralised $H_2$ production makes Carbon Capture and Storage (CCS) impractical, which is the way to reduce carbon emissions from natural gas sources. Therefore, biogas and decentralised production appear attractive regarding renewable and carbon-free $H_2$ generation even if there are additional costs regarding the removal of impurities from the biogas such as sulphur and siloxanes for reforming purposes. In addition, the dry reforming method appears particularly interesting for biogas reforming. The reactants are the major biogas compounds.

## Energy Storage and Biogas Upgrading Using Renewable Hydrogen

The renewable energy use in the overall energy mix to decrease environmental emissions and dependence on fossil fuels is mainly achieved with wind and solar energies. For example, in Denmark and Spain, wind energy has become the largest source of electricity with respective shares of 34 and 21% [29]. The main problem linked to these sources is their fluctuating and intermittent features, leading to differences between electricity supply and demand. This affects the security and stability of electrical grids [31].

To solve this issue, the flexibility of the system is a crucial point with incentive to develop flexibility through demand-side management, supply-side and infrastructure (super grids, smart grids and microgrids) flexibility [32]. The resulting mismatches between electricity production and the electricity consumption have to be balanced, and this is possible through energy storage. Different energy storage technologies have been developed. We can classify them according to their work principle:

- Mechanical energy storage (e.g. compressed air energy storage, flywheel energy storage and pumped hydro energy storage)
- Electrical energy storage (e.g. battery energy storage, flow battery energy storage, superconducting magnetic energy storage and supercapacitor energy storage)
- Thermal energy storage (e.g. sensible heat storage system and latent heat energy storage) and thermochemical energy storage (e.g. cryogenic energy storage and solar power tower)
- Chemical energy storage ($H_2$) [32–34]

All these technologies differ in term of energy capacity, response time, self-discharge, efficiency, operating constraints and their respective fields of application [35]. Overall, $H_2$ can be a good alternative for power balancing through electrochemistry. Renewable $H_2$ generation from water electrolysis using renewable electricity is a mature production method and electrolysers can reach full load operation in a few seconds enabling the $H_2$ production to adjust its power consumption quickly [36]. Therefore, $H_2$ applications as an energy storage make Power-to-$H_2$ (PtH) a way to provide temporary storage to link different sectors (electricity generation, gas grids, transport and industry) [37]; however, a current lack of infrastructure for $H_2$ storage, transport, use and the high cost of $H_2$ storage [36] are restricting.

Alternatively, $H_2$ injection into the existing natural gas network has been investigated. Regarding the infrastructure, it is reported that the transmission network, compressors, distribution network, storage components and end-use appliances allow varying degrees of $H_2$ concentration with a limiting value of 10% of $H_2$ (due to the compressors lowest tolerance). Regarding the gas quality, 10% of $H_2$ affects the supplied fuel gas calorific value under desired levels, whereas 2% of $H_2$ in the distribution network has negligible effect [38].

Since $H_2$ can be converted into $CH_4$, which is the main component of natural gas, Power-to-Methane (PtM) is promising in regions where natural gas infrastructures stand to absorb and exploit the surplus renewable energy and enable issues already developed in transport, industry and residential sectors. Moreover, PtM has issues in the power generation sector, where electricity can be produced when there is a deficit in the electricity grid due to intermittent renewable source supply [35]. Therefore, Power-to-Methane is an exciting energy storage pathway for integration of more renewable energy in the energy grid.

Power-to-Methane requires two steps: a Power-to-$H_2$ step operated through electrolysis; and, a methanation step where $H_2$ reacts with $CO_2$ to form the desired $CH_4$ [35]. Therefore, the methanation step is an exciting method from an environmental point of view, introducing $CO_2$ consumption. Moreover, in order to produce renewable methane, one has to find a renewable source of $CO_2$. Since biogas is composed mainly of $CH_4$ and $CO_2$, and considering the latter has to be removed from biogas for injection into the gas grid as upgraded biogas, the integration pathway between biogas and Power-to-Methane is a renewable energy interaction that appears as a promising issue for renewable applications.

## Methanation and Biogas Upgrading

The production of methane occurs through the Sabatier reaction (Eq. 10.14), which is the $CO_2$ hydrogeneration that produces methane from $H_2$ and $CO_2$. It is the reverse reaction of steam reforming including water-gas shift reaction (Eq. 10.13) [39].

$$CO_4 + 4H_2 \rightarrow CH_4 + 2H_2O \quad \Delta H_{298} = -164.4\ \text{kJ/mol} \qquad (10.14)$$

It is overall an exothermic reaction and is the combination of the endothermic reversed water-gas-shift (RWGS) reaction (Eq. 10.9) with the highly exothermic CO methanation (Eq. 10.15), which is the reverse methane steam reforming reaction (Eq. 10.1) [39].

$$3H_2 + CO \rightarrow CH_4 + H_2O \quad \Delta H_{298} = -206\ \text{kJ/mol} \qquad (10.15)$$

Since the highly exothermic feature of the reaction (Eq. 10.14) limits the $CO_2$ conversion at high temperature, the Sabatier reaction is favoured at relatively low temperatures where the conversion of $CO_2$ and $CH_4$ selectivity is near 100% [33]. A $CO_2$ conversion of 98% requires a temperature below 225 °C at 1 bar or 300 °C at 20 bar [27]; however, the reversed water-gas-shift reaction (Eq. 10.9) is favoured with increased temperatures [39]. Since the pressure is also a parameter that has a positive impact on efficiency, methanation is usually operated under elevated pressure [29].

## Catalytic Methanation

Kinetic limitations make the utilisation of a catalyst at relatively low temperature relevant because the reaction rate increases with temperature. The catalytic methanation requires a catalyst to be active at relatively low temperatures and enables methane selectivity. Ni-, Ru-, Rh- and Co-based catalysts may be used [29]. Due to its relatively high activity, good $CH_4$ selectivity and accessible price, nickel-based catalysts are the most used, typically on alumina support [39]; however, Ni-based catalysts have low tolerance to halogeneous and sulphurous compounds, and high-purity feed gas is required [29]. Thereby, previous cleaning processes are necessary if biogas is the $CO_2$ source used for the catalytic methanation.

A temperature range of 200–500 °C is commonly used, and a pressure range of 10–30 atm enables increased efficiency without creating high stress on the catalyst [39]. Moreover, it is reported that increased pressure increases the temperature at which carbon deposition occurs from 365 °C at 1 bar to 515 °C at 11 bar [40]. Furthermore, as it was the case with the previous methane reforming for $H_2$ production, high temperature results in carbon deposition such as through the Boudouard reaction (Eq. 10.3).

Theoretically, if the $H_2/CO_2$ ratio is higher or equal to 4 (stoichiometric ratio of Sabatier reaction), the carbon deposition should not occur [39]. Due to the high exothermic feature of the Sabatier reaction (Eq. 10.14), a critical point in the

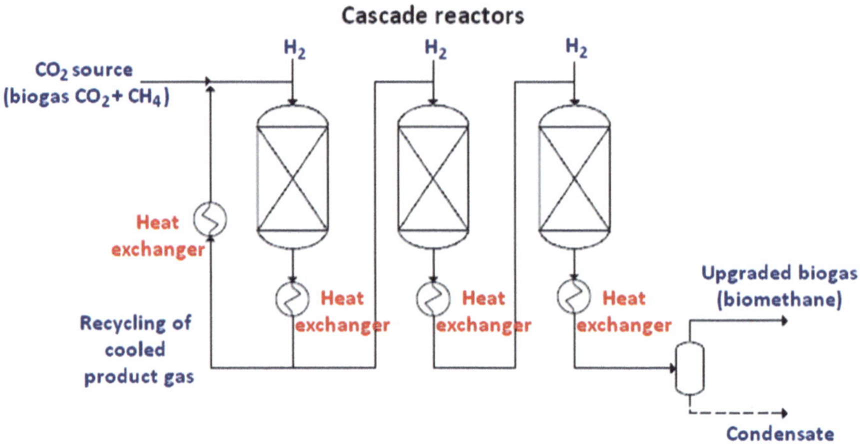


**Fig. 10.1** Simplified diagram of a series of adiabatic reactors with intercooling and cooled gas recycling

methanation process is the temperature control of the reactor to avoid thermodynamic limitations, catalyst sintering and carbon particle formation [41].

Several methanation reactors have been developed, and the main difference is the temperature profile inside the reactor [42]. The adiabatic fixed-bed reactor is usually composed of a series of adiabatic reactors with intercooling (Fig. 10.1). Recycling of the cooled product gas is commonly done to control the temperature rise in the first reactor [41]. Steam may also be added between the reactors to limit temperature increases [42]; however, temperature hot spots and pressure drops are disadvantages present with this technology. The adiabatic fixed-bed reactor is the methanation technology well developed at commercial level.

The cooled fixed-bed reactor is an alternative where the cooling operates through integrated heat exchangers such as cooling tube bundles in the reactor or cooled plates. This process is more straightforward than the adiabatic process, but the reactor itself is expensive, and hot spots are still a limitation [29].

Fluidised-bed reactors enable better control of the temperature in the reactor due to the isothermal conditions created by the mixing of fluidised solids. The heat removal with an internal heat exchanger is more effective than with fix-bed reactors, and one single reactor can be used instead of a reactor cascade. A limitation here is that $CO_2$ conversion may be incomplete due to bubbling, and the fluidisation of catalyst particles results in mechanical stress leading to catalyst loss [29, 42].

The three-phase reactor uses a liquid phase and its high heat capacity to get an effective temperature control. A fluidised bed reactor can be used, or a slurry reactor is filled with a liquid phase (such as heat transfer oil), where the catalyst particles are suspended as a result of the gas flow (Fig. 10.2). This leads to a simple process design and ideal temperature control; however, the presence of gas–liquid mass-transfer resistances is a limitation here. Moreover, the suspension liquid can decompose and evaporate [29, 42, 43].

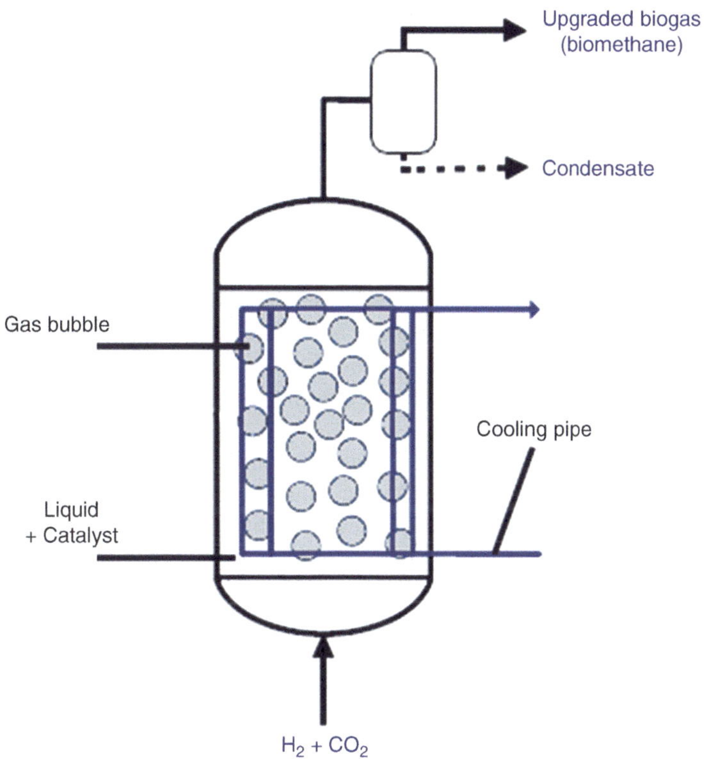


**Fig. 10.2** Simplified diagram of a slurry bubble column reactor

Structured reactors are under development to overcome the drawbacks of adiabatic fixed-bed reactors (temperature hot spots and high-pressure drops). Their internal metallic structure improves heat transport and lowers pressure drops. Microstructure reactors with high surface-to-volume ratios are also investigated. The incorporation of the catalyst onto the structure and its replacement when it is deactivated are two limitations [44].

Sorption-enhanced methanation reactors are another development of structured fixed-bed reactors. The water produced during the reaction is adsorbed on the catalyst support, leading to a reduction in the thermodynamic limit of $CO_2$ conversion. Temperature or pressure swing methods can then be used to remove the water and regenerate the support [42].

## Biological Methanation

In biological methanation, methanogenic microorganisms in a liquid phase play the role of biocatalysts. This direct production of methane is done through hydrogenotrophic methanogenesis from $CO_2$ and $H_2$. The activity of methanogens occurs at a temperature range from 0 to 122 °C, and optimal temperatures are comprised between 15 and 98 °C [45]. The temperature of anaerobic digestion has no impact on the methane yield; however, it has a significant impact on the reaction velocity and response time of the biological methanation.

Since hydrogenotrophic methanogenesis also represents the last step in the production of biogas, two techniques are possible for biological methanation: ex situ methanation in a separate reactor and in situ methanation in a biogas digester (Fig. 10.3):

- **Ex situ**: Here, the $H_2$ and $CO_2$ gases are injected in a separate reactor in a stoichiometric $H_2/CO_2$ ratio of 4 and converted by microorganisms. This process is called ex situ methanation. The microorganisms can come from pure cultures of methanogen or a mixed culture obtained from sewage sludge or a working biogas plant [45].
- **In situ**: $CO_2$ reduction through $H_2$ injection in the digester enables increased yields of biogas with a higher calorific value due to the conversion of $CO_2$ to methane. This method enables upgrading of the biogas directly in the digester and would reduce the cost of post upgrading process and the need for other plants for injection into the natural gas grid as biomethane. Despite this, the technique is not well developed yet and still faces some challenges. For example, due to the additional injection of $H_2$ in the digester, there is an increase of the $H_2$ partial pressure. This leads to an undesirable environment for some bacteria, which can be inhibited and lead to lower production of acetate. Additionally, the injection of $H_2$ in situ increases the pH of the environment. It has been found that the production of acetate and methane is favoured with an $H_2$ partial pressure lower than 745 mbar [45].

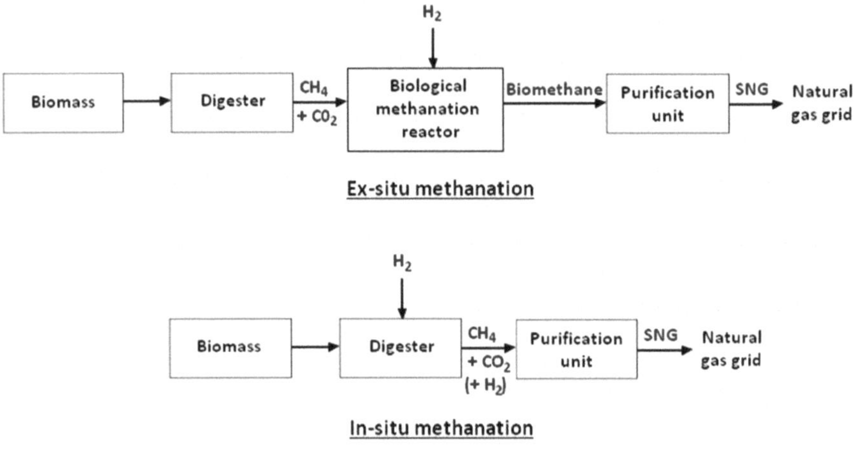


**Fig. 10.3** Process flow diagram for ex situ and in situ biological methanation

In order to reduce investment costs, efficient reactors are required. For biological methanation, three reactor systems that can be feed with $H_2/CO_2$ mixtures:

- Thoroughly stirred tank reactors, where the heat distribution is homogeneous and mixing favours the up flow of gas bubbles
- Fixed-bed reactors (anaerobic filters) that promote conversion with thin layers of biomass that increases the surface-to-volume ratio
- Trickle bed reactors that are based on the fixed-bed reactor concept where the reactor chamber is not filled with liquid, but the nutrient liquid is continuously wetting the surface of microorganisms

Stirred tank reactors and fixed-bed reactors are the only concepts that can be used in situ because trickle bed reactor cannot produce its $CO_2$ and be fed only with $H_2$. In situ methanation requires fewer investment costs than ex situ by using existing biogas plants, but ex situ biological methanation avoids several challenges that occur with in situ concept. Additionally, the methane conversion rate is much higher with ex situ compared to in situ process.

## Methanation of Biogas and Comparison of Methanation Technologies

Using biogas directly as the feed gas for methanation reduces the yield of the reaction and requires an excess of $H_2$ [46]. Despite this, the integration of methanation in the biogas chain enables increases in its specific methane yield. Among the different methanation technologies, catalytic methanation is already used at commercial level, whereas biological methanation is developed only at pilot plant scale [45].

Since the reactor size is a crucial aspect regarding investment, the required reactor volume to produce the required gas quality and methane volumetric flow can be compared for different methanation processes through their Gas Hourly Space Velocity (GHSV). GHSV is defined as the volumetric flow rate of the feed gas $\dot{Q}_{\text{feed gas}}$ without any inert gases and with a stoichiometric $H_2/CO_2$ ratio per reactor volume $V_R$ (Eq. 10.16).

$$\text{GHSV} = \frac{\dot{Q}_{\text{feed gas}}}{V_R} \tag{10.16}$$

If we assume the conversion leading to a methane content higher than 90% in the output gas, the catalytic adiabatic fixed-bed methanation has the higher GHSV leading to the smaller reactor volume required to get the required product gas, compared to catalytic three-phase methanation and biological methanation in a separate stirred tank reactor [29]. The GHSV values and significant features for these three technologies are reported in Table 10.2.

**Table 10.2** Comparison of biological and catalytic methanation processes [29, 45]

| Parameter/ feature | Biological methanation (in a separate reactor) | Catalytic methanation (adiabatic fixed-bed reactor) | Catalytic methanation (three-phase bed reactor) |
|---|---|---|---|
| GHSV ($h^{-1}$) | <<100 | 2000–5000 | 500–1000 |
| T (°C) | 15–98 | 300–550 | 300–350 |
| Phases | G, L, (S) | G, S | G, L, S |
| Back mixing | High | Low | Moderate |

Three significant features explain these differences of performance: (1) process temperature, since a higher reaction temperature leads to a higher reaction rate; (2) mass transfer phenomena, since the presence of a liquid phase leads to gas–liquid mass transfer resistances; and, (3) back mixing of the gas and liquid phase that has a strong influence at high conversion rates.

Biological methanation operates at a lower temperature and has a lower reaction rate. The presence of the liquid phase leads to extra mass transfer resistances, and liquid and gas phases that are assumed perfectly mixed lead to a high back mixing. These features explained the low GHSV value for biological methanation compared to catalytic methanation that operates at higher temperatures [29].

The presence of a liquid phase in catalytic three-phase methanation compared to the adiabatic fixed-bed, and the moderate back mixing in this liquid phase added to a lower temperature of reaction explains why the GHSV is lower in three-phase methanation than in the adiabatic fixed-bed.

Biological methanation does have higher tolerance towards impurities of the feed gas [45]. Moreover, the absence of a nickel catalyst that can be poisoned with sulphur compounds present in biogas means that biological methanation does not require its previous purification. Additionally, carbon deposition and catalyst deactivation are avoided with biological methanation. Regarding the power requirement, biological methanation requires more power than catalytic methanation because of the need for a stirrer to introduce $H_2$ effectively into the liquid phase, consuming around 10% of the input energy [29].

## Power-to-Biogas Process

A typical power-to-biogas plant is characterised by at least four parts (Fig. 10.4):

- An electrolyser, which allows $H_2$ to be produced from renewable electricity
- A source of $CO_2$, which can be biogas
- A methanation reactor, where the $H_2$ and $CO_2$ are converted into methane
- Storage facilities and compressors

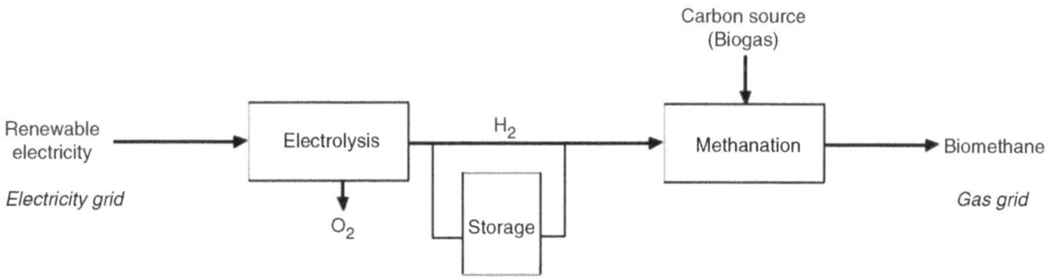


**Fig. 10.4** Process flow diagram for power-to-biogas

## Carbon Source

Regarding the carbon source, $CO_2$ is mainly used from CCS (Carbon Capture and Storage), biomass, industrial processes as a by-product and air [46]. Biogas plants are carbon sources that can help to store several MW of chemical energy via the PtG process chain [29]. Therefore, they appear as a suitable carbon source to store the electricity surplus. $CO_2$ from biogas can be added in the methanation step after removal from the biogas upgrading step, but the direct use of biogas in the methanation step and a higher interaction between biomass-to-gas and power-to-gas processes is an exciting issue leading to biogas upgrading without additional energy or cost effort [35].

## Electrolysers Consideration in the Case of PtG Chain

Among the electrolyser technologies, alkaline, PEM and SOEC are the leading technologies considered for Power-to-Gas utilisation. Their principal features are summarised in Table 10.3. The alkaline is the most mature and cheapest electrolysis technology with widespread commercialisation. PEM electrolysis is in an early period of commercialisation and is more expensive in terms of initial investment costs but seems to be more suitable for PtG plants due to its ability to absorb intermittent energy offering better operating flexibility [35]. Furthermore, a drawback of alkaline electrolysers is its limited operating load range, whereas flexibility is an efficiency criterion for PtG and renewable energy storage. Despite this, new

**Table 10.3** Summary of major features of electrolyser [29, 34, 35, 47]

| Parameter/feature | Alkaline | PEM | SOEC | MCEC |
|---|---|---|---|---|
| Maturity | Mature | Early phase of commercialisation | Mature | Development |
| Exchanged ions | $OH^-$ | $H^+$ | $O^{2-}$ | $CO_3^{2-}$ |
| T (°C) | 60–80 | 50–80 | 600–1000 | 700–900 |
| Efficiency (%) | 62–82 | 67–82 | 90 | 90 |
| Cold start time | 15–20 min | <15 min | >60 min | – |

alkaline electrolysers would allow variable operation ranging from 5 to 100% of the nominal capacity and are operational within only a few minutes.

SOEC operates at much high temperatures. There are still some challenges due to this high operational temperature, and the cold start-up time is also a drawback of SOEC for flexible application; however, the high temperature accelerates the kinetics and reduces the reversible cell voltage of the electrolysis process leading to higher efficiencies [47].

Molten Carbonate Electrolysis Cells (MCEC) have a temperature range and efficiency similar to SOEC and could also be in the future an option for integration in the PtG chain but are not yet developed enough for currently possible utilisation.

## The Efficiency of the Process

The PtG process has an average efficiency of 54%, if an average efficiency of the electrolysis of 70% and an average methanation efficiency of 78% are considered (without taking into account the heat utilisation) [35]. The combination of biomass-to-gas and power-to-gas technologies has many advantages, including low gas cleaning expenses and the possibility to increase system efficiency [29].

Regarding the global PtG chain, the use of the heat released by the exothermic methanation is necessary to increase its efficiency. With catalytic methanation that operates at a higher temperature than biological methanation, the wasted heat is highly valuable for power production, district heating or utilisation of steam. Biological methanation and its operating temperature lead to fewer opportunities for integration of the waste heat. However, the integration with a biomass-to-gas chain enables valorisation of waste heat with low exergy content as heat source for the biogas digester. Additionally, the $O_2$ from the electrolysis can be used for primary desulphurisation of biogas into the digester [29]. Moreover, in order to reduce water consumption, the water produced through methanation is recycled to the electrolyser.

## Economic Consideration

Regarding the investment costs, electrolysers are the most crucial expenditure covering 74% of the total costs, 21% of this total expenditure is used for $H_2$ storage, whereas 2.5% and 1.8% are respectively spent on compression and methanation [29]. This shows the relevance of flexible methanation to lower the $H_2$ storage need and its relative cost.

One advantage of this process is to consider the methanation from the PtG chain as an upgrading technique for biogas. Regarding the production costs for upgrading biogas, a study based in Spain shows that production costs are more than halved with variable upgraded methane production instead of continuum production [48], showing that it is more relevant to store biogas and operate upgrading processes when cheap electricity surplus is present through the PtG chain.

Generation costs, including valorisation of heat and $O_2$, are comprised between 0.16 and 0.39 €/kWh with an expectation of 0.07–0.10 €/kWh by 2050 [29]. Even if Power-to-Biogas is not yet competitive with natural gas generation (0.03 €/kWh) or with direct biomethane generation (0.07 €/kWh) [29], this concept is promising since it has vast benefits regarding energy storage possibilities, $CO_2$ consumption, biogas upgrading and integration of renewable energies.

## Conclusions

To conclude, chemical conversion possibilities between $H_2$ and methane lead to promising integration pathways for renewable energy applications. The conversion of methane to $H_2$ through reforming techniques enables a waste-to-$H_2$ pathway, providing a second life to residues and renewable production. Moreover, the conversion of $H_2$ to methane through methanation techniques results in a power-to-biomethane pathway, which enables power balancing between different sectors and empowers intermittent renewable energy source penetration into the overall energy mix.

**Acknowledgements** The authors are grateful to the ENERSENSE programme and NTNU Team Hydrogen at the Norwegian University of Science and Technology (NTNU) for supporting and helping on this book project.

## References

1. Scarlat N, Dallemand J-F, Fahl F (2018) Biogas: developments and perspectives in Europe. Renew Energy 129(A):457–472
2. Sarker S, Lamb JJ, Hjelme DR, Lien KM (2018) Overview of recent progress towards in-situ biogas upgradation techniques. Fuel 226:686–697
3. Sarker S, Lamb JJ, Hjelme DR, Lien KM (2019) A review of the role of critical parameters in the design and operation of biogas production plants. Appl Sci 9(9):1915
4. Lamb JJ, Sarker S, Hjelme DR, Lien KM (2018) Fermentative bioethanol production using enzymatically hydrolysed *Saccharina latissima*. Adv Microbiol 8:378
5. Lamb JJ, Hjelme DR, Lien KM (2019) Carbohydrate yield and biomethane potential from enzymatically hydrolysed Saccharina latissima and its industrial potential. Adv Microbiol 9 (4):359–371
6. Gaoa Y et al (2018) A review of recent developments in hydrogen production via biogas dry reforming. Energy Convers Manag 171:133–155
7. Patinvoh RJ et al (2017) Innovative pretreatment strategies for biogas production. Bioresour Technol 224:13–24
8. Alves HJ et al (2013) Overview of hydrogen production technologies from biogas. Int J Hydrogen Energy 38:5215
9. Sun Q et al (2015) Selection of appropriate biogas upgrading technology—a review of biogas cleaning, upgrading and utilization. Renew Sustain Energy Rev 51:521–532
10. Ryckebosch E, Drouillon M, Vervaeren H (2011) Techniques for transformation of biogas to biomethane. Biomass Bioenergy 35:1633–1645
11. Surendra KC et al (2014) Biogas as a sustainable energy source for developing countries: opportunities and challenges. Renew Sustain Energy Rev 31:846–859

12. Khan MI, Yasmin T, Shakoor A (2015) Technical overview of compressed natural gas (CNG) as a transportation fuel. Renew Sustain Energy Rev 51:785–797
13. Shanmugam K, Tysklind M, Upadhyayula VKK (2018) Use of liquefied biomethane (LBM) as a vehicle fuel for road freight transportation: a case study evaluating environmental performance of using LBM for operation of tractor trailers. Proc CIRP 69:517–522
14. Budzianowski WM (2016) A review of potential innovations for production, conditioning and utilization of biogas with multiple-criteria assessment. Renew Sustain Energy Rev 54:1148–1171
15. Nikolaidis P, Poullikkas A (2017) A comparative overview of hydrogen production processes. Renew Sustain Energy Rev 67:597–611
16. Acar C, Dincer I (2019) Review and evaluation of hydrogen production options for better environment. J Cleaner Prod 218(1):835–849
17. Di Marcoberardino G et al (2018) Potentiality of a biogas membrane reformer for decentralized hydrogen production. Chem Eng Process Process Intensif 129:131–141
18. Nawfal M (2015) Valorisation catalytique du biogaz pour une énergie propre et renouvelable. Catalyse
19. Nahar G, Mote D, Dupont V (2017) Hydrogen production from reforming of biogas: review of technological advances and an Indian perspective. Renew Sustain Energy Rev 76:1032–1052
20. Cruz PL et al (2018) Exergy analysis of hydrogen production via biogas dry reforming. Int J Hydrogen Energy. https://doi.org/10.1016/j.ijhydene.2018.02.025
21. Noh YS, Lee K-Y, Moon DJ (2019) Hydrogen production by steam reforming of methane over nickel based structured catalysts supported on calcium aluminate modified SiC. Int J Hydrogen Energy 44(38):21010–21019
22. Rau F et al (2018) Efficiency of a pilot-plant for the autothermal reforming of biogas. Int J Hydrogen Energy. https://doi.org/10.1016/j.ijhydene.2018.04.014
23. Battista F et al (2017) LCA evaluation for the hydrogen production from biogas through the innovative BioRobur project concept. Int J Hydrogen Energy 42:14030–14043
24. Braga LB et al (2013) Hydrogen production by biogas steam reforming: a technical, economic and ecological analysis. Renew Sustain Energy Rev 28:166–173
25. Durán P et al (2019) Pure hydrogen from biogas: intensified methane dry reforming in a two-zone fluidized bed reactor using permselective membranes. Chem Eng J 370:772–781
26. Fernandez E et al (2015) Development of thin Pd-Ag supported membranes for fluidized bed membrane reactors including WGS related gases. Int J Hydrogen Energy 40(8):3506–3519
27. Leonzio G (2019) ANOVA analysis of an integrated membrane reactor for hydrogen production by methane steam reforming. Int J Hydrogen Energy 44(23):11535–11545
28. Junaedi C et al (2012) Development of integrated reformer systems for syngas production. Int J Hydrogen Energy 37(13):10435–10443
29. Götz M et al (2016) Renewable power-to-gas: a technological and economic review. Renew Energy 85:1371–1390
30. Enerdata, Global Energy Statistical Yearbook 2018. Accessed 10 June 2019. https://yearbook.enerdata.net
31. Bailera M et al (2017) Power to gas projects review: lab, pilot and demo plants for storing renewable energy and CO2. Renew Sustain Energy Rev 69:292–312
32. Lund PD et al (2015) Review of energy system flexibility measures to enable high levels of variable renewable electricity. Renew Sustain Energy Rev 45:785–807
33. Mahlia TMI et al (2014) A review of available methods and development on energy storage; technology update. Renew Sustain Energy Rev 33:532–545
34. Burheim OS (2017) Engineering energy storage. Elsevier, London
35. Ghaib K, Ben-Fares F-Z (2018) Power-to-methane: a state-of-the-art review. Renew Sustain Energy Rev 81:433–446
36. Pääkkönen A, Tolvanen H, Rintala J (2018) Techno-economic analysis of a power to biogas system operated based on fluctuating electricity price. Renew Energy 117:166–174

37. Dolci F et al (2019) Incentives and legal barriers for power-to-hydrogen pathways: an international snapshot. Int J Hydrogen. https://doi.org/10.1016/j.ijhydene.2019.03.045
38. Gondal IA (2019) Hydrogen integration in power-to-gas networks. Int J Hydrogen Energy 44 (3):1803–1815
39. Stangeland K et al (2017) CO2 methanation: the effect of catalysts and reaction conditions. Energy Proc 105:2022–2027
40. Jürgensen L et al (2015) Dynamic biogas upgrading based on the Sabatier process: thermodynamic and dynamic process simulation. Bioresour Technol 178:323–329
41. Er-rbib H, Bouallou C (2014) Methanation catalytic reactor. Comptes Rendus Chimie 17 (7–8):701–706
42. Rönsch S et al (2016) Review on methanation—from fundamentals to current projects. Fuel 166:276–296
43. Lefebvre J et al (2015) Improvement of three-phase methanation reactor performance for steady-state and transient operation. Fuel Process Technol 132:83–90
44. Kreitz B, Wehinger GD, Turek T (2019) Dynamic simulation of the CO2 methanation in a micro-structured fixed-bed reactor. Chem Eng Sci 195:541–552
45. Lecker B et al (2017) Biological hydrogen methanation—a review. Bioresour Technol 245 (Pt A):1220–1228
46. Mazza A, Bompard E, Chicco G (2018) Applications of power to gas technologies in emerging electrical systems. Renew Sustain Energy Rev 92:794–806
47. David M, Ocampo-Martínez C, Sánchez-Peña R (2019) Advances in alkaline water electrolyzers: a review. J. Energy Storage 23:392–403
48. Curto D, Martín M (2019) Renewable based biogas upgrading. J Cleaner Prod 224:50–59

# Lifetime Expectancy of Lithium-Ion Batteries

# 11

Lena Spitthoff, Jacob J. Lamb, Bruno G. Pollet, and Odne S. Burheim

## Introduction

The current efforts of replacing fossil fuels and traditional energy sources through the exploitation of renewable energy have led to a massive increase in the lithium-ion battery (LIB) market. LIBs have become the leading energy storage technology in many sectors due to their high gravimetric and volumetric energy, high power density, high efficiency, low self-discharge property and compatibility with the existing electric infrastructure [1]. The transport sector has a significant impact on the increase in the LIB market, mainly where hybridisation is beneficial or where operation time is shorter than charging time. The sales of electric vehicles were accelerating and reached a total of million sales in July 2018. This includes battery electric passenger vehicles (BEV), plug-in hybrid electric vehicles (PHEV) and e-buses [2] with LIBs being the established energy storage technology [3].

Moreover, the marine transport sector has an increasing interest in LIBs with the first all-electric car and passenger ferry launched in Norway in 2015 and increasing

L. Spitthoff
Department of Energy and Process Engineering, ENERSENSE, Norwegian University of Science and Technology, Trondheim, Norway
e-mail: lena.spitthoff@ntnu.no

J. J. Lamb (✉)
Department of Electronic Systems, ENERSENSE, Norwegian University of Science and Technology, Trondheim, Norway

Department of Energy and Process Engineering, ENERSENSE, Norwegian University of Science and Technology, Trondheim, Norway
e-mail: jacob.j.lamb@ntnu.no

B. G. Pollet · O. S. Burheim
Department of Energy and Process Engineering, ENERSENSE and NTNU Team Hydrogen, Norwegian University of Science and Technology, Trondheim, Norway
e-mail: bruno.g.pollet@ntnu.no; odne.s.burheim@ntnu.no



J. J. Lamb, B. G. Pollet (eds.), *Micro-Optics and Energy*,
https://doi.org/10.1007/978-3-030-43676-6_11

numbers [4, 5]. The hybridisation of cruise ships and supply ships is also growing significantly. Another growing application for LIBs is the grid support when introducing higher shares of intermittent renewable energies. AES Corporation launched a solar and battery plant with a capacity of 100 MWh storage capacity of LIB in January 2019 [6].

Moreover, the use in off-grid power supply systems, e.g. to support ferry charging as local grid often have low power capacity, receives growing attention. However, for a successful application of LIBs in these market segments and to keep up with the traditional or alternative systems, there are still obstacles to overcome. The most urgent requirements are the production cost reduction, accompanied by an increase in battery life.

This chapter aims to give a brief introduction into the working principle of LIBs, the most common commercially available chemistries and summarises the most recent developments in applications of LIB and individual new requirements. Moreover, different studies investigating the capacity degradation are compared in terms of the influence of operating conditions on the lifetime of LIBs.

## Terminology

### Working Principle of an LIB

An LIB cell (as schematically shown in Fig. 11.1) utilises two open-structured materials as electrodes, where the lithium (Li) ions intercalate. The positive and negative electrode is attached to current collectors and separated by a separator. An electrolyte provides Li-ions and prevents the electrons from flowing through. While charging, the material of the positive electrode dissolves in the electrolyte solution, forming a positive ion and an electron in an oxidation reaction [7]. The Li-ions move to the negative electrode through the electrolyte while the electrons circulate through the external circuit. The negative electrode consumes the electrons upon intercalation of the Li-ions. During discharging, the Li-ions and electrons circulate in opposite directions than during charging. Per definition the electrode that has the oxidation reaction during discharge is the anode and the reduction reaction takes place at the cathode [8]. As LIBs are operated outside the electrochemical stability range of the electrolyte, the organic electrolyte is reduced and forms a passivating layer on the anode. This layer behaves similarly to solid electrolyte interphase and is called SEI (solid electrolyte interphase) layer. It mostly prevents further reduction of electrolyte; however, an unstable SEI layer leads to further consumption of Li-ions and increase in the impedance of the cell [9].

### Applications, Requirements and Problems

The application of an LIB prescribes important parameters like power requirements and life expectancies and therefore determines the selection of the battery system. On

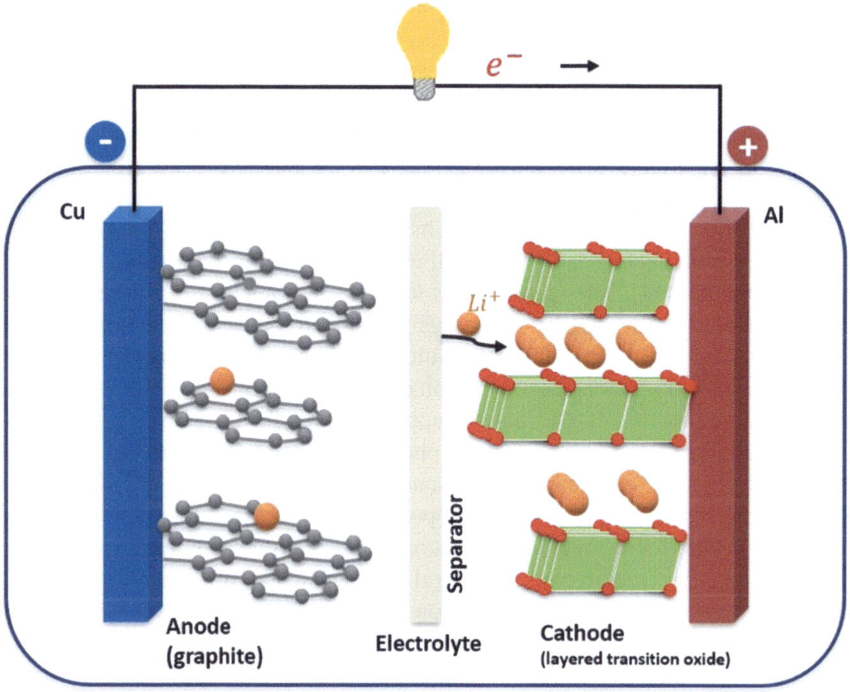


**Fig. 11.1** Schematic illustration of an Li-ion battery under discharge. The Li-ions are moving from the anode to the cathode while the electrons circulate through the external circuit

the other hand, it also affects the operating conditions of the battery like the state of charge (SOC) window, temperature range, and charging and discharging rate which impacts the ageing of the battery.

For the application in the transport sector, the power and energy requirements, the power to energy ratios or C rates are highly depending on the utilisation within the vehicle. The C rate is the rate at which a battery is discharged relative to its initial capacity. A 1 C rate means that the discharge current will discharge the battery in 1 h.

Electric vehicles (EV) are commonly divided into battery electric vehicle (BEV), plug-in hybrid electric vehicle (PHEV) and hybrid electric vehicles (HEV). BEV runs entirely on a battery and electric drive train. PHEV uses a battery and an internal combustion engine to recharge the battery or replace the electric drive. HEV has a gasoline engine and fuel tank plus an electric motor and a battery. They use no charging from the grid but get their energy from gasoline and regenerative braking [10]. While the required power and energy is the highest for EV (power 50–350 kW, energy 30–90 kWh) followed by PHEV (power 40–150 kW, energy 5–15 kWh), is the required C rate the largest for HEV with around 4 C, followed by PHEV with around 2 C and BEV uses a C rate around 0.5 C. The battery in an electrical vehicle is commonly used in the SOC window of 10–100% or 20–100%. For all vehicle applications, low weight is crucial.

In addition, fast charging, low weight and low costs are essential for PHEV and EV applications [11]. The specific requirements in the marine transport depend on the transport type (e.g. general marine transport, international shipping and small vessels) and are around 1 MWh and 0.8–1.5 C in the example of the all-electric car ferry [5, 12, 13]. For the application of LIBs in aviation, the EU published milestones for 2050. This contains, for example, hybridisation of 30–50% for power and 10–20% for stored energy as well as full electric flights for more than 500 km for personal aviation [14]. The requirements for aerospace applications depend on the specific application for the LIB (e.g. usage for onboard power or complete electrification including take-off and landing resulting in high power requirements).

Another market sector is the application of LIBs in combination with the electricity grid as well as for off-grid power supply. Combining LIBs with the grid will support the integration of high shares of photovoltaic and wind energy in the power mix by providing storage capacity and ancillary services. The usage of LIBs for off-grid power offers, for example, the possibility of providing reliable access to electricity in developing regions when combining with solar home systems. The power and energy requirements for grid-connected and off-grid power supply systems depend on the specific application with some grid application needing a C rate of up to 10 C (e.g. solar home system requires energy around 0.1–10 kWh) [15]. The main requirements for this type of application is a long calendar and cycle lifetime with up to 20 years [16, 17] and over 2500 full cycles [3]. Moreover, for LIBs to be able to compete with other systems, a cost reduction is necessary [17].

## Second Life

The end of life of an LIB used in EVs is reached when the capacity drops below 80% of the initial capacity [11]. The end of life for this LIB application was defined to ensure the satisfaction of driver mobility even though recent studies question the definition and propose more flexible criteria to define the end of life [18]. Even though the LIBs are rejected in the transport sector, they are still capable of providing energy storage for applications that are less critical in terms of battery performance, volume and weight. According to Martinez-Laserna et al. [19], many second-life demonstration projects, mainly reusing EV batteries, currently demonstrate that those batteries are still capable of fulfilling the demands in stationary applications. They also have pointed out in their extensive review that different studies have suggested market prices ranging from 44 to 180 $/kWh for retro filled modules. When considering the costs of the refurbishment of EV batteries, accounting for the collection to selling, the EV batteries are responsible for around half of the costs. The most significant additional refurbishment expenses are labour, general and administrative expenses, and packaging costs. Martinez-Laserna et al. [19] also have pointed out that it is still unclear if a second life will provide any discount for EV owners and if the most profitable stationary applications will provide enough market to allocate a large number of batteries expected to be retired from the transport sector in the upcoming years. They criticise the lack of studies on ageing performance and safety of second-life batteries even though it is crucial for the economic and technical viability as well as the environmental impact mitigation.

Additionally, they concluded that only batteries that have not reached the ageing knee (change in dominant ageing mechanism) should be considered for a second-life application. Casals et al. [20] calculated the lifespan of second-life batteries using an equivalent electric battery-ageing model and pointed out the strong lifespan dependency on battery use. The life expectancy varies from around 30 years in fast electric vehicle charging support applications to around 6 years in community energy storage systems.

## Safety Concerns and Ageing

The critical requirements for the successful application of LIBs in segments like the transport sector are high energy and high power, which lead to more safety issues of LIBs [21]. These safety issues become even more critical when considering LIBs for ships as incidents on the sea brings a higher impact and even higher for application of LIBs for aerospace applications. This was elucidated when in 2013 Boeing 787 Dreamliners worldwide were grounded due to safety concerns with the LIBs used for onboard backup power during flight [22]. Further, cargo planes and EV buses that have caught fire caused by LIBs are reported [23]. When exposed to abusive conditions, like mechanical abuse (crush, nail penetration, etc.), electrochemical abuse (e.g., overcharge and short-circuiting) and thermal abuse (e.g., external heating and flame attack), self-sustained cascaded exothermic reactions may be triggered which can evolve into a thermal runaway and subsequent rupture and violent fires.

As already mentioned, one of the main requirements to successfully use LIBs for many applications is long battery life. Therefore, the degradation of battery performance due to different ageing mechanisms is frequently studied for different cell chemistries. During the usage of batteries, many factors affect the ageing, such as (ambient) temperature, current, accumulated ampere-hour throughput and SOC windows [24–28]. Battery degradation is known to be a result of several simultaneous physicochemical processes. These processes are commonly divided into three categories: loss of Li inventory (LLI), loss of active materials (LAM) and impedance increase due to reaction kinetics degradation [29]. LLI is mainly caused by the formation of the SEI layer, Li plating and different Li consumption rates at the cathode and anode. LAM can be introduced by the dissolution of material, structural degradation, particle isolation and electrode delamination. Increased impedance is a result of the formation of passive films at the active particle surfaces, pore clogging as well as electrolyte conductivity losses [24]. The ageing and underlying mechanisms are investigated in different ways. Dubarry et al. [24] categorised the methods roughly into experimental, modelling, post-mortem and electrochemical voltage spectroscopy. Experimental ageing studies on different commercially available chemistries will be compared in terms of ageing below.

## Definitions and Calculations

Ageing characteristics can be presented in different ways. One way of evaluating the battery degradation is relating to the cell coulombic capacity at a specific point to the initial cell capacity. This relative cell capacity is frequently used in ageing studies [26, 27] and mainly referred to as the state of health (SOH) [30], normalised capacity [28] and relative, remaining or relative discharge capacity [17, 25, 31]. Another way of evaluating battery degradation is by presenting the accumulated capacity loss normalised to initial conditions using the terms maximum available capacity degradation [32] and degradation % age [33]. Ageing studies can generally be divided into cycle life and storage (shelf, calendar) life studies. Storage life studies look at battery degradation versus the time that a battery was stored under specific conditions. For this, the reviewed calendar ageing studies performed characterisation tests before storing and periodically after storing at the defined SOC and temperature for specific time intervals. The characterisation tests mostly consist of capacity measurements using 1–3 cycles and IR measurements at room temperature (defined either at 20 °C or 25 °C).

Cycle life studies look at battery degradation when continuously cycled. For the evaluation, the amount of cycling is quantified in different ways. Some studies look at the number of cycles defined as one charge and one discharge until specified cut-off voltages [29, 32, 33] or the degradation to the time that the battery is cycled [28, 30]. Other studies relate the current cycle capacity to the original cell capacity. This method is applied using the term normalised cycles [25], full equivalent cycles [27] or equivalent full cycles [29]. This method will be referred to as full equivalent cycles (FEC) below. The cycling time and number of cycles were calculated to FEC to compare ageing data from different studies. The cycling time was converted to FEC by multiplying with half of the C rate as the C rate is the rate at which a battery is discharged relative to its initial capacity. A 1C rate means that the discharge current will discharge the battery in 1 h. The number of cycles was converted to FEC by multiplying the cycle number with the depth of discharge (e.g. 80% in case of a SOC window of 10–90%) and integrating the capacity loss over the number of cycles using a linear regression between two measurement points. When evaluating degradation characteristics batteries are cycled in different SOC windows (e.g. 20–80% SOC) or stored at a specific SOC. Some studies use the term Depth of Discharge (DOD) instead of SOC. The DOD can either be given from 100% SOC (e.g. 60% DOD is equal to a SOC window of 40–100%), or around 50% SOC (e.g. 60% DOD around 50% is equal to a SOC window of 20–80% SOC) [27].

## Chemistries

The commercially available LIBs are commonly composed of a carbon-based intercalation anode and a cathode consisting of mixed lithium transition metal oxides. The most frequently used cathode materials are lithium cobalt oxide (LCO), nickel cobalt manganese oxide (NMC), nickel cobalt aluminium oxide (NCA), lithium manganese oxide (LMO) and lithium iron phosphate (LFP) [34]. These cathode and anode compounds are generally mixed with small amounts of conductive agents (e.g. carbon black) and a polymeric binder (e.g. polyvinylidene fluoride) and tape cast onto a metal current collector (aluminium for the cathode and copper for the anode). Brief introductions into the cathode materials LCO, LMO, NMC and LFP and hard carbon and graphite as anode material, the most commonly used separator materials and electrolyte solutions are given below. Various reviews by Nitta et al. [34], Fergus et al. [35], Liu et al. [36] and Marom et al. [1] give a good and more detailed overview of the most commonly used chemistries in LIBs.

## Cathode Materials: LCO, LMO, NMC and LFP

Goodenough first introduced $LiCoO_2$ (LCO) in 1980 [37]. It became the first successful LIB on the market (Sony combining LCO cathode with carbon anode) and is still one of the most used cathode materials of commercially available LIBs. [36] It is a layered transition metal oxide. Cobalt (Co) and Li which are located in octahedral sites occupy alternating layers and form a hexagonal symmetry. The paths for lithiation and delithiation are planes. The average operation voltage is 3.8 V [34]. Despite a high theoretical capacity of 270 mAh/g, only around 145 Ah/kg can be reversibly accessed due to the risk of $O_2$ evolution. The main drawback with LCO is the toxic and expensive Co and poor thermal stability. New materials are being developed trying to replace Co with cheaper transition metals like manganese, nickel and iron to obtain cheaper materials and increase Li utilisation [35].

By substituting Co in LCO cathodes with Manganese (Mn), the cathode material $LiMn_2O_4$ (LMO) was obtained. It has a spinal structure with manganese occupying the octahedral sites and Li predominantly occupying the tetrahedral sites [38]. Here, the paths for lithiation and delithiation are instead a three-dimensional network of channels [39]. The average operating voltage is 4.1 V and the theoretically available specific capacity 148 Ah/kg [34].

The cathode material $LiNi_{0.5}Mn_{0.5}O_2$ (NMO) was the result of extensive research efforts to create a cheaper cathode material. However, to enhance rate capability, Co was added and resulted in the formulation of $LiNi_xCo_yMn_zO_2$ (NMC). NMC has a similar or higher achievable specific capacity than LCO while having lower costs due to a reduced Co content. The most commonly used form is $LiNi_{1/3}Co_{1/3}Mn_{1/3}O_2$ with an average voltage of 3.7 V and theoretically available specific capacity of 280 Ah/kg [34]. However, compositions with increased Ni content recently received an increase in interest as they provide larger specific capacities and energy densities and lower the costs due to the higher capacity and lower costs of Ni compared to Co

**Table 11.1** Application vs. chemistries [3, 16, 17, 34, 40–42]

| Chemistry | Cell cost ($/kWh) | Theoretic-specific capacity (Ah/kg) | Energy density (Wh/kg) | Application |
|---|---|---|---|---|
| LCO | 200 | 274 | 190 | Majority of commercial LIBs |
| NMC333 | 250 | 280 | 160 | Dominating in EV and PHEV applications; projections for grid-connected uses; tailoring for special applications, portable electronics, power tools and medical devices |
| LMO | 170 | 148 | 150 | e-bikes; power tools and medical devices |
| LFP | 170 | 170 | 150 | Marginal role in EV, e-bikes; huge potential for use in power supply systems both off-grid and grid-connected |

[43]. The common abbreviation NMC is added with numbers indicating the decimals of each element (e.g. NMC523: $LiNi_{0.5}Co_{0.2}Mn_{0.3}O_2$). Nickel is associated with higher capacity, cobalt with better rate capability and manganese with improved safety [44]. Goodenough first reported $LiFePO_4$ (LFP) in 1996 [37]. It has an olivine structure with covalently bonded $PO_4$ groups forming the $PO_4$ tetrahedron together with a chemically stable $Fe^{2+}/Fe^{3+}$ couple within the $FeO_6$ octahedron. During charging and discharging, the Li-ions are respectively extracted from and inserted into $LiFePO_4$ while the central iron ions are oxidised and reduced. The main advantages of this material are the use of inexpensive and environmentally benign Fe as well as the excellent safety characteristics due to the structure and voltage plateau at 3.4 V which suppresses electrolyte degradation. The main disadvantage is the low volumetric energy density with an available theoretical capacity of 170 Ah/kg [34]. Table 11.1 gives an overview of the current and predicted applications of the presented chemistries.

## Anode Materials: Hard Carbon and Graphite

Anode materials are commonly divided into three main categories, depending on their Li-ion battery performances and reaction mechanisms. These are intercalation anodes which are mostly carbon based; alloy anodes like silicon and germanium; and conversion anodes using transition metal compounds like oxides, phosphides, sulphides and nitrides [45].

The commercial anode material of choice is carbon as it combines very desirable properties like low cost, low delithiation potential versus $Li/Li^+$, high Li diffusivity, high electrical conductivity and relatively low volume changes during lithiation and delithiation. Carbon anode materials can be divided into graphitic carbons and hard carbons [34]. Graphite is a three-dimensional ordered crystal with large graphite

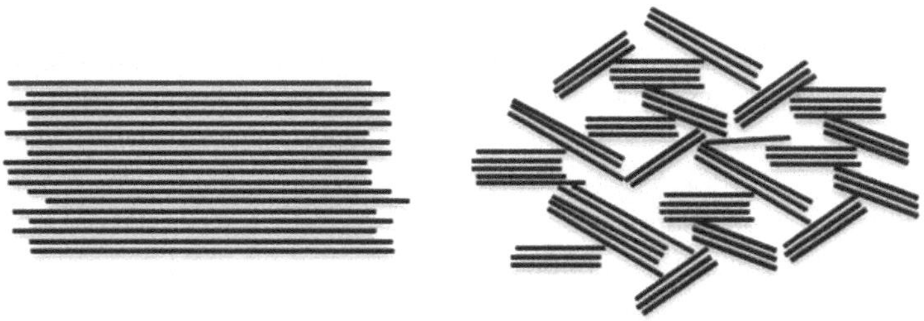

**Fig. 11.2** Structure of graphite and hard carbon

grains with a density of 2.26 g/cm$^3$ [46]. The surface most exposed to the electrolyte is the basal plane, but the transport of the Li-ions take place in the cross-section plane of the graphite particles [47]. During the intercalation of Li, the graphitic particles undergo uniaxial a strain along the edge planes [48]. These strains may damage the SEI and reduce the cell's cycle life. Hard carbon or "non-graphitizing carbon" was never graphitised and consists of two dimensional "ordered" graphene sheets which are randomly stacked (Fig. 11.2). Its density is between 1.4 and 1.8 g/cm$^3$ and is smaller than graphite [46]. Peled et al. [47] showed that the SEI on hard carbon is thicker and explained it by the accumulation of electrolyte-reduction products in surface voids. Moreover, hard carbon has a higher fraction of exposed edge planes and therefore an increased quantity of SEI formed in the first cycles [34].

## Separator, Electrolyte and Additives

The purpose of the separator is preventing direct contact between the cathode and the anode and thus avoiding short-circuiting. Commercially used separators utilise mostly polyolefin-based microporous membranes like polyethylene (PE), polypropylene (PP) or PE/PP composite membranes. They have excellent mechanical properties, chemical stability and economic. However, they show poor safety performance and their low porosity, insufficient electrolyte wettability and low melting point are affecting the electrochemical properties of LIBs [49]. In recent years, research was focused on improving the performance of polyolefin-based separators by coating conventional membranes. These coated separators showed improved thermal stability [50–52]. However, this method reduces the porosity, and more recent approaches have been focusing on using electrospinning to obtain microporous nanofiber membranes with high porosity and enhanced wetting properties [49, 53, 54].

The electrolyte is interacting with all components in the battery and is, therefore, determining power density, cycling stability and safety of the battery. The most important criteria for a suitable electrolyte is an ideal ionic conductivity while being electronically insulating, a full electrochemical stable window to prevent electrolyte degradation, being inert to other components as well as low costs, non-toxic, and sustainable. Most of the electrolytes used in commercially available LIBs are non-aqueous solutions utilising Lithium hexafluorophosphate ($LiPF_6$) salt, dissolved in an organic carbonate, mostly mixtures of ethylene carbonate (EC), dimethyl carbonate (DMC), propylene carbonate (PC), diethyl carbonate (DEC) and ethyl methyl carbonate (EMC) [55]. Electrolyte additives are used to improve the performance of LIBs. The amount is usually no more than 5%. Additives are mostly utilised to improve the SEI formation, protect the cathode, stabilise the $LiPF_6$ salt and increase safety [56].

## Capacity Fading and Ageing Characteristics

In this section, capacity fading and ageing characteristics are compared for both calendar ageing and cycling ageing for different chemistries and studies. This is done by comparing the FEC required to reach a certain SOH at different temperatures, DOD and charging rates for the cycling studies and the storage time required to reach a certain SOH at different temperatures and SOC for the calendar ageing studies. The amount of FEC and the storage time at specific SOH were obtained by linearly interpolating between the two closest measurement values. It was as far as possible attempted to use data that was as comparable as possible, meaning that parameters that are not changed are constant at the same values. Sometimes this was not possible as different conditions were chosen for the ageing studies or different parameter were varied. In these cases, the most similar data sets were chosen. To our understanding the selected comparisons are fair, at least on the first order of magnitude.

### Capacity Fade as a Function of Temperature, C Rate and SOC Window

Results from the cycle life studies for the cathode materials NMC, NMC/LMO, LCO and LFP are presented in Fig. 11.3 as SOH versus FEC. All studies were performed with a charging and discharging rate of 1 C and within a SOC window of 0–100% and a temperature of 25 °C ("LFP" 30 °C).

We observe a reasonable and similar cycling behaviour for NMC and LFP until reaching 3000 FEC. With further cycling, the NMC follows a steady decrease in SOH until the end of the study with nearly 6000 FEC. A SOH of 80% and thus the defined end of life for automotive applications was reached after nearly 4000 FEC. The LFP has a drop-in SOH after around 3000 FEC but appears to stabilise after this drop with a faster degradation rate than before. An SOH of 80% is reached after around 3150 FEC. The SOH of the NMC/LMO blend and the LCO, on the other hand, decrease with a more significant rate. The SOH of the NMC/LMO blend performs similar to the SOH of the LFP and NMC for the first 500 FEC before it decreases quickly, reaching an SOH of 80% around 800 FEC. The SOH of the LCO starts decreasing quickly from the start, reaching a SOH of 80% already around 500 FEC. The study on the NMC performed by Richter et al. [25] analysed the degradation behaviour of a 17.5 Ah NMC/hard carbon battery and the study on the

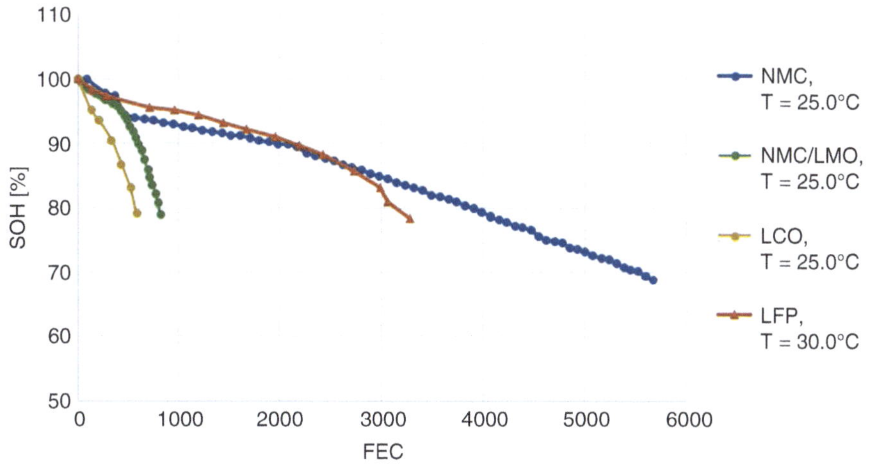


**Fig. 11.3** SOH as a function of FEC. All batteries were cycled the same conditions (charging and discharging rate of 1 C and within a SOC window of 0–100%) except from the difference in temperature ("NMC" [25], "NMC/LMO" [30] and "LCO" [32]: 25 °C; "LFP" [27]: 30 °C) (Anode: "NMC/LMO": graphite and carbon particles, "NMC": hard carbon, "LCO" and "LFP": graphite)

LFP performed by Grolleau et al. [57] analysed a 15 Ah graphite/LFP battery, whereas all the other evaluated studies were done with batteries in the range of 1.5–3 Ah.

Figure 11.4 shows the FEC relative to the number of cycles at 25 °C (18 °C for LFP) at constant SOH. It illustrates the accelerating effect of increased temperature on batteries using NMC/LMO blend, NMC, LCO, and LFP as the cathode material. Although NMC and LFP age slowest, it appears that they are more temperature sensitive compared to LCO and NMC/LMO. When considering a temperature increase from 25 to 55 °C for the NMC battery, the battery can only be cycled 5% of the FEC before realising an SOH of 80%. It is also striking that except for an SOH of 95% the temperature dependency is nearly linear and very similar for different SOH, which is vital for temperature accelerated ageing tests. The capacity degradation of the battery with NMC/LMO shows a lower but still significant temperature dependency with a decreasing impact with ageing. The lowest impact can be observed for the battery with the LCO cathode. Increasing the temperature from 25 to 55 °C results into a 60% decrease of FEC at an SOH of 95%. The battery with LFP cathode shows a decrease in capacity with temperature in the range of the battery with NMC cathode. It is also noticeable that NMC, LCO, LFP have the most significant temperature dependency in a temperature interval between 25 and 40 °C. A further increase in temperature has a declining effect on capacity degradation. The temperature dependence of the NMC/LMO shows the same trend for an SOH of 95% but appears to be most sensitive to temperature changes in a temperature interval of 50–60 °C for further cycling.

Figure 11.5 illustrates the accelerated capacity fading due to an increased charging rate. Generally, the chemistries with the most inferior ageing properties are also

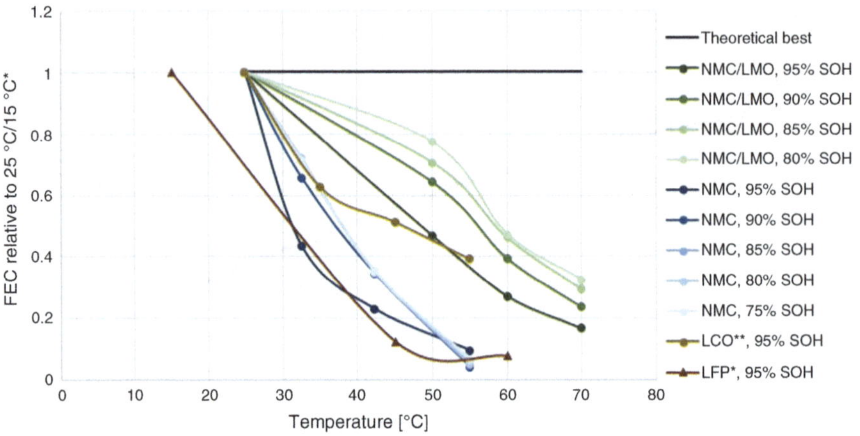


**Fig. 11.4** Cycling ageing as a function of temperature. "NMC" [25] and "NMC/LMO" [30] related to 25 °C, charging and discharging rate of 1 C and 100% DOD, "LCO" [33] similar but charging rate of 0.5 C instead, "LFP" [28] related to 18 °C, discharge rate 0.5 C and data summarised for DOD of 10–90%. (Anode: "NMC/LMO": graphite and carbon particles, "NMC": hard carbon, "LCO" and "LFP": graphite)

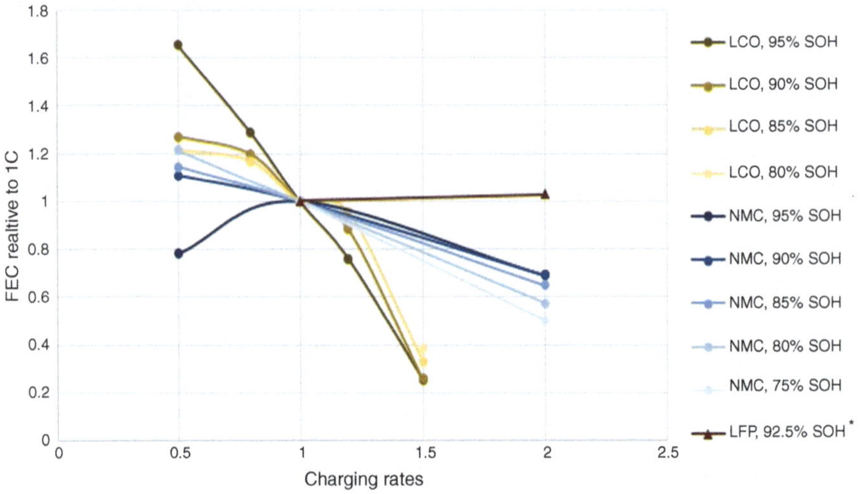


**Fig. 11.5** Cycling ageing as a function of the charging rate. "NMC" [25] and "LCO" [32] at 25 °C and fixed discharging rate at 1 C, "LFP" [27] 30 °C, the discharge rate is discharging similar to charging rate

the ones most sensitive to rate changes. The effect is most pronounced for LCO. Only 30% of the FEC can be realised until a SOH of 85% when increasing the charging rate from 0.5 to 1.5 C. This effect is even more significant at the beginning of cycling. For the battery with NMC cathode, the effect of the charging rate

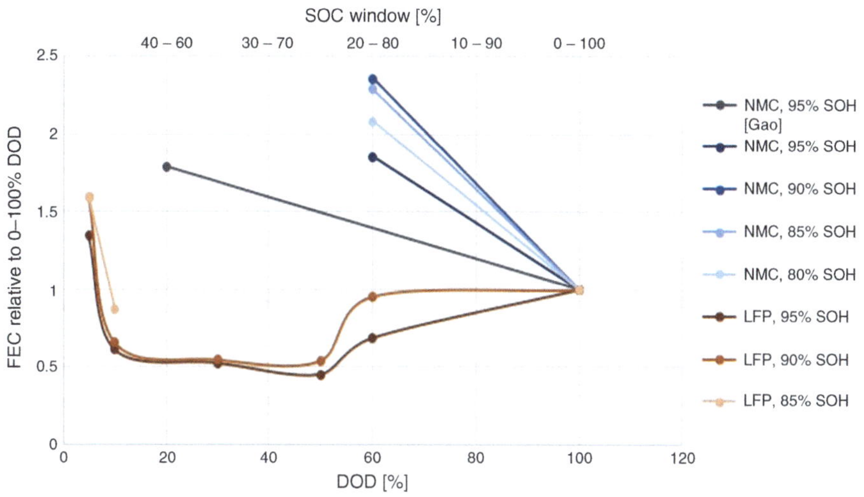


**Fig. 11.6** Cycling ageing as a function of DOD around 50% SOC (e.g. 60% DOD is equal to a SOC window of 20–80%). "NMC (Gao)" [29] at 25 °C and 6 C charging and discharging rate, "NMC" [25] at 25 °C and 2 C charging and discharging rate and "LFP" [27] at 30 °C and 1 C charging and discharging rate. (Anode: "NMC": hard carbon, "NMC (Gao)" and "LFP": graphite)

increases with ageing. Only around 50% of the FEC can be realised before reaching an SOH of 80% when increasing the charging rate from 0.5 to 1.5 C. It is noticeable that while dependency on charging rate for NMC seems to be linear, the dependency for the LCO is higher when charging rates of 1 C and higher are reached. For the battery with LFP cathode, the increase in charging and discharging rate seems to have nearly no effect.

Figure 11.6 shows the amount of FEC for different SOC windows around an SOC of 50% relative an SOC window of 0–100%, which was the only common point of reference. The batteries with NMC cathode appear to benefit from decreasing the SOC window. When decreasing the SOC window from 0–100% to 20–80%, nearly two times the amount of FEC can be realised before an SOH of 95% is reached. A further decrease to 40–60% seems to bring no further improvement. The battery with LFP cathode shows a complicated relationship between SOC window and capacity degradation with plateaus between the DOD of 100–60% and 50–10%. Decreasing the SOC window from 0–100% to 25–75% bisects the amount of FEC that can be realised until a specific SOH. A further decrease in the SOC window until 45–55% has no additional effect. Overall the battery with LFP cathode seems to benefit from full cycles.

One can summarise that NMC and LPF perform best at a constant temperature but are very sensitive to an increase in temperature. The temperature has the most significant impact on the ageing behaviour for all the chemistries except the LCO, which is mostly impacted by an increased charging rate. The effect on an increased charging rate and SOC window seem to be in the same order of magnitude for the NMC with the effect of the SOC window being smaller for the study using a more significant variation in SOC window and a graphite anode instead of hard carbon, which leaves it open if a further reduction of the SOC window is accelerating ageing again or if the anode material is responsible for the change in behaviour. The battery with LFP cathode seems to be most sensitive to an increase in temperature, followed by the variation in SOC window and showing nearly no effect when changing the charging rate.

Except from the extensive ageing study on NMC/hard carbon performed by Richter et al. [25] most studies were not performed until a sufficiently low SOH was reached to draw any conclusion about the utilisation for second-life applications. However, the results on NMC/hard carbon battery look promising for a charging and discharging rate of 1 C and temperatures of 25 and 32.5 °C.

For both data series, the ageing knee is not reached in the study after cycling for around 1500 FEC after reaching the SOH of 80%. For cycling at elevated temperature (42.3 °C, 55 °C), the ageing knee was reached at an SOH between 50 and 55% with approximately the same amount of FEC until 80% SOH and between 80% SOH and the ageing knee. Only for an increased discharge rate to 2 C the ageing knee was reached earlier and even before an SOH of 80% which was prevented when additionally decreasing the SOC window to 20–80%.

## Calendar Ageing: Capacity Fade as a Function of SOC and Temperature

Results from the calendar life studies for the cathode materials NMC and LFP are presented in Fig. 11.7 as SOH versus storage time. All batteries were stored at a SOC of 100%, and a temperature 20 and 30 °C. We observe similar behaviour for the battery with NMC cathode and graphite anode and the two batteries with LFP cathode. The battery with NMC cathode and hard carbon anode, however, has the largest decrease in SOH with time. The SOH appears to have a negative logarithmic dependency on the storage time. Figures 11.8 and 11.9 illustrate the considerably accelerated capacity degradation during storage with an increase in temperature and SOC.

Figure 11.8 shows the storage time relative to a temperature between 40 and 45 °C for batteries with NMC cathode and hard carbon anode and LFP cathode and a graphite anode. The effect of increased temperature is more severe for the NMC cathode and increases with ageing. An SOH of 85% is reached 6 times faster when increasing the storage temperature from 25 to 42.5 °C and 15 times faster when increasing the storage temperature to 55 °C. The three compared studies on batteries with LFP cathode all show that the capacity degradation is increasing with temperature as well but less intense than for the NMC cathode. An SOH of 90% is reached 1.3 times faster when increasing the storage temperature from 25 to 40 °C and five times faster when increasing the storage temperature to 60 °C.

In Fig. 11.9, we observe the storage time relative to an SOC 90 or 100% for batteries with NMC cathode and hard carbon anode and LFP cathode and graphite anode. The effect of an increased SOC during storage is more severe for the NMC

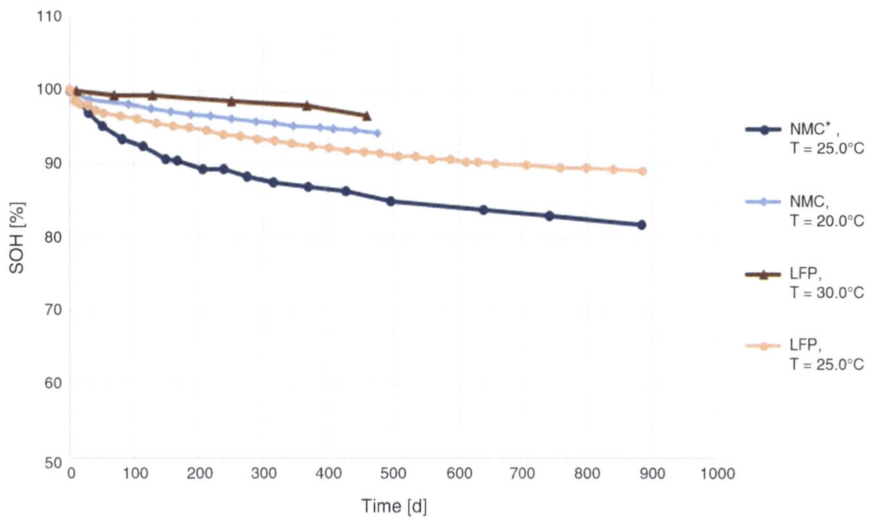


**Fig. 11.7** SOH as a functional storage time in days. All batteries were stored at an SOC of 100% and temperatures of 20 °C ("NMC" [31]), 25 °C ("NMC*" [25], "LFP" [17]) and 30 °C ("LFP" [57]) (graphite anode, except NMC*: hard carbon)

cathode and is increasing with ageing. The effect of an increased SOC appears to be linear in the SOC window from 60 to 90% before it reaches a plateau for a further increase until 100%. A SOH of 90% would be reached seven times faster when storing a battery with NMC at 100% instead of 60%. The effect of an increased SOC during storage is accelerating the capacity degradation less severe for batteries with LFP cathode.

One can summarise for the calendar ageing both increased SOC and temperature accelerate capacity degradation with the temperature being damaging. These trends appear the same for both batteries with NMC cathode with hard carbon anode and LFP cathode with graphite anode but are more severe for the NMC battery. While the NMC/hard carbon battery performed best when evaluating the cycling, the LFP/graphite batteries are more stable in terms of calendar ageing. An SOH of 90% was reached for an LFP cathode and graphite anode after 260 days at 40 °C and an SOC of 90% in [26], after around 220 days at an SOC of 100% (25 °C: 660 days) in [17] and around 100 days at an SOC of 100% and a temperature of 45 °C in [57]. The same SOH was already reached after around 70 days for an NMC cathode and hard carbon anode in [25] after storage at 80% SOC and 42.3 °C (25 °C: 360 days).

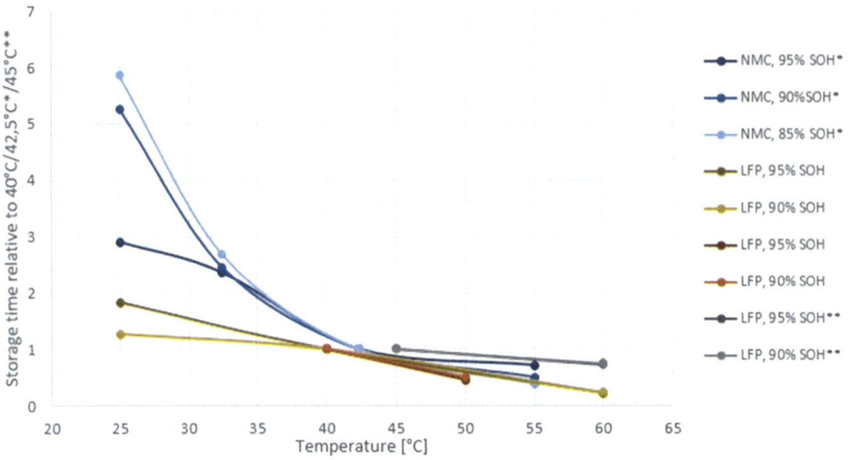


**Fig. 11.8** Calendar ageing as a function of temperature. Series marked with asterisk are related to 42.5 °C instead of 40 °C and series marked with double asterisk are related to 45 °C instead of 40 °C. Anode material: "NMC" [25]: hard carbon, "LFP" [17, 26, 57]: graphite

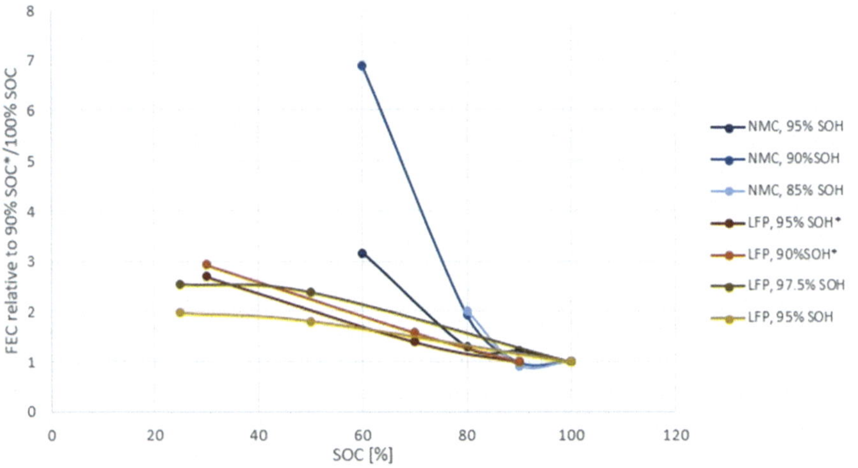


**Fig. 11.9** Calendar ageing as a function of SOC. Series marked with asterisk are related to 90% SOC instead of 100% SOC. Anode material: "NMC" [25]: 25 °C, hard carbon, "LFP" [26] [31]:40 °C, graphite

## Ageing Mechanisms

Ageing mechanisms are strongly dependent on the electrode compositions and differ for the cathode and the anode. As both anode and cathode materials, as well as other parameters (e.g. battery capacity, format, etc.), differ for the evaluated degradation studies it is difficult to assign the observed degradation characteristics to specific ageing mechanisms.

The primary ageing mechanism on carbon-based anodes is the development of the SEI layer over time which consumes Li-ions. This is more severe at a higher SOC (>80%) and accelerated with temperature [30]. A high SOC results in a low anode potential which causes electrochemical instability on the electrolyte material [27]. A further expansion of the SEI layer can be triggered during storage at high voltage. Solvent interaction with graphite after diffusion of solvent components through the SEI layer can induce graphite exfoliation and gas creation leads to cracking of the SEI layer which again leads to further SEI growth. Additional Li-ions are consumed due to side reactions on the anode [58]. High temperatures can furthermore lead to a dissolving of the SEI and creation of Li salts less permeable to Li-ions which results in an increased anode ohmic impedance. At low temperatures a decrease of Li diffusivity within SEI and the carbon anode can lead to plating of metallic Li and subsequent reaction with the electrolyte, leading to loss of cyclable Li. The temperature at which plating of Li happens depends on the polarisation potential versus $Li/Li^+$ and is, therefore, material dependent [30].

In present electrolyte solutions with $LiPF_6$, trace HF is often inevitable and leads to the dissolution of transition metal ions. This results in a reduction in active mass and passivation of the negative carbon electrode due to deposition of this transition metal compounds. This leads to increased thermal resistance and is accelerated with increased temperature. Aurbach et al. [59] showed that this degradation mechanism is more critical for LCO and LFP but showed a high stability $LiMn_{1.5}Ni_{0.5}O_4$ and $LiMn_{0.5}Ni_{0.5}O_2$ and attributed it to the unique surface chemistry due to the presence of nickel ions in the lattice.

Waldmann et al. [30] showed structural changes of the LMO/NMC composite cathode depending on temperature with the changes of NMC lattice constants correlating to the loss of Li-ions at the cathode. Naumann et al. [17] showed that the loss of active material on the negative electrode when using an LFP cathode at elevated temperatures were the result of Fe deposition on the anode.

While the capacity fading of the LFP/graphite battery with temperature was in the range of the NMC/hard carbon battery, it was far more stable during storage. Many studies demonstrated the excellent stability performance of LFP cathodes which is in agreement with the observed behaviour during storage. A possible explanation for the improved stability of the NMC/hard carbon battery compared to the LFP/graphite batteries during cycling might be the difference in anode material. Graphite is exposed to strain along the edge plains due to expansion during the lithiation which can lead to cracking of the SEI, thus resulting in freshly exposed graphite and therefore a further degradation of electrolyte and SEI layer formation. Hard carbon, on the other hand, has an initially thicker SEI layer but also a more stable one.

### Data Extraction Sensitivity

When evaluating the extracted ageing data, two sensitive points have to be kept in mind. First, the quality of the figure inserted into the *ScanIt* software is crucial as the software uses pixels to calculate data from the set axes and data points. Second, the selecting of data points is supported by the software, but sometimes overlapping data sets make it difficult to pick the correct data point.

### Comparison Justification

When relating the ageing data to a specific data set it was tried to use the same conditions. Sometimes this was not possible, and the most similar data sets were chosen as reference (e.g. when relating to temperatures (40 °C/42.5 °C/45 °C) were chosen). Additionally, when comparing the impact of charging rate data was chosen with only a change in charging rate, but for one data set, the discharging rate is changed accordingly. All the differences were noted while presenting the data. Moreover, the size, geometry, and other components are different for the compared studies. Therefore, no quantitative comparison between the studies was made; only dependencies and ageing characteristics were examined.

## Conclusions

Demands on energy/power density, cost reduction, lifetime and safety are steadily increasing and are major driving forces in the LIB research. The common strategies on increasing the energy density on the material level are to improve the specific and volumetric capacities as well as their operating potentials (i.e. maximising the cell voltage with cathode materials operating at high electrode potentials and anode materials operating at low electrode potentials). The most promising trends for the next years in the LIB world are electrodes with a significant higher theoretical capacity such as silicon [60], increasing energy density by using higher cell voltages [61] and the utilisation of solid-state electrolytes to enable higher charging rates [62] and improved safety [63].

Silicon anodes are generally seen as the next generation of anodes for LIB with an up to 10 times higher gravimetric capacity than currently used carbon anodes and a decreased potential [60]. However, silicon anodes show poor cycling stability due to significant volume expansion (300–400%) and contraction and cracking during the lithiation and delithiation. During this volume changes the SEI layer is damaged, and therefore, new SEI is formed continuously and consumes Li-ions. Different strategies to achieve long-term capacity retention are currently explored. A significant focus is on using nanostructured silicon to overcome surface cracking, and particle fractures as Liu et al. [64] found that the critical particle size is 150 nm. Nanomaterials and structures like nanoparticle, nanowires, nanotubes and nanoporous networks have shown superior electrochemical properties [65–68], but industrial applications are still challenging as their fabrication is not yet cost-effective [45]. Other efforts are focused on finding new binder materials [69, 70], new electrolytes and electrolyte additives (e.g. Fluoroethylene carbonate) that stabilises the electrolyte/electrode interphase [69, 71] or surface modification like molecular layer deposition [69].

Other considered options are the coating of silicon particle to obtain an artificial SEI layer [69, 72], increasing the porosity [73] and using composite anodes (e.g. formed by silicon, Li and highly electronically conductive materials such as TiC/C) [66, 74]. Moreover, applying special cycling procedures that use for example limited potential windows [75–77] or prelithiation of the anode by introducing an alternative Li source [78–81] are considered.

One approach to improve the energy and power densities of LIBs is to increase the cell voltage by either decreasing the anode potential or increasing the cathode potential. Relevant cathode materials include nickel-rich layered oxides ($LiNi_{1-x}M_xO_2$, M = Co, Mn, and Al), Li-rich layered oxides ($Li_{1+x}M_{1-x}O_2$, M = Mn, Ni and Co), high-voltage spinel oxides ($LiNi_{0.5}5Mn1_{1.5}O_4$), and high voltage polyanionic compounds (e.g. phosphates, sulphates and silicates) [61]. This chapter showed the detrimental impact of a high SOC due to a high potential on the ageing behaviour of different battery chemistries. This is even more severe when considering high voltage batteries as some of these materials have their operating voltages almost entirely outside of the thermodynamic stability window of the electrolyte solutions. Li et al. [61] summarised the main degradation mechanisms for high voltage cathode materials. Some of these mechanisms appear to be known ageing mechanisms which are increased in severity for high voltage like the dissolution of active positive electrode material due to thermal and hydraulic decomposition of $LiPF_6$ and surface structural reconstruction and mechanical fracture of the positive electrode material and chemical crossover of species between the two electrodes. Additional degradation mechanisms are the parasitic electrolyte oxidation at the charged positive electrodes resulting in a passivation layer on the cathode similar to the SEI on graphite, spontaneous reactions involving conductive carbon additives due to anodic instability at high voltages, instability of other cell components at extreme potentials and exothermic breakdown of interphases on both electrodes.

The technologically most advanced option for high voltage batteries with improved gravimetric energy density, where currently much research is focusing on, is NMC cathode materials with increased nickel content ($LiNi_xCo_yMn_zO_2$ with $x > 0.6$) [44, 82–86]. One of the significant drawbacks with nickel-rich NMC is decreased thermal stability. Some EVs of the current generation are already employing NCM with a nickel:cobalt:manganese ratio of 5:3:2 [44], which was shown to have so far the optimised composition offering a reasonably good thermal stability and high capacity due to a well-balanced ratio of nickel content to manganese and cobalt contents [87]. The main problems with nickel-rich NMC cathode, summarised by Schipper et al. [44], are the large amount of highly reactive $Ni^{4+}$ at high SOC that lead to undesired side reactions with the electrolyte solution, the poor thermal stability resulting in serious safety concerns, capacity fade due to Li and

Nickel cation mixing and therefore spinel formation on the surface. Additionally, from a ratio of 8:1:1 cycling will lead to cracks which result in increased surface area and thus more active sites for parasitic reactions. Research is focused on ensuring stability while increasing the nickel content. Research is mainly focused on creating artificial cathode–electrolyte interphases either by introducing functional groups on the surface [84] or electrolyte additives [86].

Solid-state electrolytes (SSE) have attracted considerable attention lately as one of the critical components for all-solid-state LIBs allowing high safety and increased energy density. By replacing the organic liquid electrolytes and separator with an inorganic solid electrolyte it is possible to use higher potential cathode materials due to improved electrochemical stability. As SSE are not organic, they are not flammable. Additionally, SSEs are reported to enable higher charging rates as in contrast to liquid electrolytes no concentration gradient is build up during charging and discharging at high rates [62]. The major problem with SSE is about ten times lower conductivity compared to liquid electrolytes resulting in high resistance and therefore high losses at high rates [63]. Different electrolytes are currently investigated with NASICON-, garnet- and perovskite-type solid electrolytes being the most promising ones for application in all-solid-state LIBs. However, Zheng et al. [63] concluded in their extensive review that further improvement is necessary for the compatibility between solid electrolytes and electrodes.

**Acknowledgements** The authors are grateful to the ENERSENSE programme and NTNU Team Hydrogen at the Norwegian University of Science and Technology (NTNU) for supporting and helping on this book project.

## References

1. Marom R, Amalraj SF, Leifer N, Jacob D, Aurbach D (2010) A review of advanced and practical lithium battery materials. J Mater Chem 21:9938–9954
2. Bloomberg NEF (2018) Cumulative global EV sales hit 4 million. https://about.bnef.com/blog/cumulative-global-ev-sales-hit-4-million/. Visited 20 May 2019
3. Zubi G, Dufo-Lopez R, Carvalho M, Pasaoglu G (2018) The lithium-ion battery: state of the art and future perspectives. Renew Sust Energ Rev 89:292–308
4. C Shahan (2015) World's first all-electric battery-powered ferry. Clean Technica. https://cleantechnica.com/2015/06/13/worlds-first-electric-batterypowered-ferry/. Accessed 07 May 2019
5. Corvus Energy World's first all-electric car ferry. Corvus Energy. https://corvusenergy.com/marine-project/mf-ampere-ferry/. Accessed 07 May 2019
6. Spector J (2019) AES completes record-breaking solar and battery plant on Kauai. Greentech Media. https://www.greentechmedia.com/articles/read/aes-completes-its-recordbreaking-solar-and-battery-plant-on-kauai#gs.gst76r. Accessed 07 June 2019
7. Julien C, Mauger A, Vijh A, Zaghib K (2016) Lithium batteries. Springer, Cham
8. Rahn C, Wang C (2013) Battery system engineering. Wiley, West Sussex
9. Besenhard J (2011) Handbook of battery materials. Wiley-VCH, Weinheim
10. Ogura K, Kohle M (2017) Battery technologies for electric vehicles. In: Electric vehicles: prospects and challenges. Elsevier, Cambridge, pp 139–167

11. Perner A, Vetter J (2015) Lithium-ion batteries for hybrid electric vehicles and battery electric vehicles. In: Advances in battery technologies for electric vehicles. Elsevier, Amsterdam
12. Corvus Energy AT6500 high performance energy storage. Corvus Energy. http://files7.webydo.com/42/421998/UploadedFiles/15e14716-e3fd4bd4-af37-6a67a86f74a5.pdf. Accessed 07 May 2019
13. Siemens Norge Snart blir trafikken over norske fjorder elektrisk. Siemens Norge. https://w3.siemens.no/home/no/no/topics/fremtidenelektrisk/pages/elektrisk-bilferge.aspx. Accessed 07 May 2019
14. Affenzeller J, Beaumel L, Bergstein M, Coppin O, Faye I, Hildermeier J, Isikveren A, Perlo P, Pfluger J, Scott D, Thulin N (2017) Electrification of the transport system: studies and reports. European Commission
15. Ansean D, Gonzalez M, Garcia V, Viera J, Anton J, Blanco C (2015) Evaluation of $LiFePO_4$ batteries for electric vehicle applications. IEEE Trans Ind Appl 51:1855–1863
16. Naumann M, Karl R, Truong C, Jossen A, Hesse H (2015) Lithium-ion battery cost analysis in PV-household application. Energy Procedia 73:37–47
17. Naumann M, Schimpe M, Keil P, Jossen A, Hesse H (2018) Analysis and modeling of calendar aging of a commercial $LiFePO_4$/graphite cell. J Energy Storage 17:153–169
18. Saxena S, Le Floch C, MacDonald J, Moura S (2015) Quantifying EV battery end-of-life through analysis of travel needs with vehicle powertrain models. J Power Sources 282:265–276
19. Martinez-Laserna E, Gandiaga I, Sarasketa-Zabala E, Stroe D-I, Swierczynski M, Goikoetxea A (2018) Battery second life: hype, hope or reality? A critical review on the state of art. Renew Sustain Energy Rev 93:701–718
20. Casals L, García B, Canal C (2019) Second life batteries lifespan: rest of useful life and environmental analysis. J Environ Manag 232:354–363
21. Goodenough J, Kim Y (2010) Challenges for rechargeable Li batteries †. Chem Mater 22:587–603
22. Williard N, He W, Hendricks C, Pecht M (2013) Lessons learned from the 787 Dreamliner issue on Lithium-ion battery reliability. Energies 6:4682–4695
23. Wang Q, Ping P, Zhao X, Chu G, Sun J, Chen C (2012) Thermal runaway caused fire and explosion of lithium ion battery. J Power Sources 208:210–224
24. Dubarry M, Qin N, Brooker P (2018) Calendar aging of commercial Li-ion cells of different chemistries - a review. Curr Opin Electrochem 9:106–113
25. Richter F, Vie P, Kjelstrup S, Burheim O (2017) Measurements of ageing and thermal conductivity in a secondary NMC-hard carbon Li-ion battery and the impact on internal temperature profiles. Electrochim Acta 250:228–237
26. Sarasketa-Zabala E, Gandiaga I, Rodriguez-Martinez LM, Villarreal I (2014) Calendar ageing analysis of a $LiFePO_4$/graphite cell with dynamic model validations: towards realistic lifetime predictions. J Power Sources 272:45–57
27. Sarasketa-Zabala E, Gandiaga I, Rodriguez-Martinez L, Villarreal I (2015) Cycling ageing analysis of a $LiFePO_4$/graphite cell with dynamic model validations: towards realistic lifetime predictions. J Power Sources 275:573–587
28. Wang J, Liu P, Hicks-Garner J, Sherman E, Soukiazian S, Verbrugge M, Tataria H, Musser J, Finamore P (2011) Cycle-life model for graphite$LiFePO_4$ cells. J Power Sources 196:3942–3948
29. Gao Y, Jiang J, Zhang C, Zhang W, Jiang Y (2018) Aging mechanisms under different state-of-charge ranges and the multi-indicators system of state-of-health for lithium-ion battery with Li (NiMnCo)O2 cathode. J Power Sources 400:641–651
30. Waldmann T, Wilka M, Kasper M, Fleischhammer M, Wohlfahrt Mehrens M (2014) Temperature dependent ageing mechanisms in Lithium-ion batteries - a post-mortem study. J Power Sources 262:129–135
31. Schmitt J, Maheshwari A, Heck M, Lux S, Vetter M (2017) Impedance change and capacity fade of lithium nickel manganese cobalt oxide-based batteries during calendar aging. J Power Sources 353:183–194

32. Gao Y, Jiang J, Zhang C, Zhang W, Ma Z, Jiang Y (2017) Lithium-ion battery aging mechanisms and life model under different charging stresses. J Power Sources 356:103–114
33. Leng F, Tan C, Ming C, Pecht M (2015) Effect of temperature on the aging rate of Li ion battery operating above room temperature. Sci Rep 5:12967
34. Nitta N, Wu F, Lee J, Yushin G (2015) Li-ion battery materials: present and future. Mater Today 18:252–264
35. Fergus J (2010) Recent developments in cathode materials for lithium ion batteries. J Power Sources 195:939–954
36. Liu F, Mukherjee P (2015) Materials for positive electrodes in rechargeable lithium-ion batteries. In: Rechargeable lithium batteries. Elsevier, Cambridge, pp 21–39
37. Mizushima K, Jones PC, Wiseman PJ, Goodenough JB (1980) $Li_xCoO_2$ ($0<x\leq1$): a new cathode material for batteries of high energy density. Mater Res Bull 15:783–789
38. Thackeray M, Kock A, David W (1993) Synthesis and structural characterization of defect spinels in the lithium-manganese-oxide system. Mater Res Bull 28:1041–1049
39. Whittingham M (2004) Lithium batteries and cathode materials. Chem Rev 104:4271–4302
40. Yoo G-W, Jang B-C, Son J-T (2015) Novel design of core shell structure by NCA modification on NCM cathode material to enhance capacity and cycle life for lithium secondary battery. Ceram Int 41:1913–1916
41. Gallagher K, Dees D, Nelson P (2015) PHEV battery cost assessment. Vehicle technologies program annual merit review and peer evaluation meeting, argonne national laboratory, May 9–13th
42. Burheim OS (2017) Engineering energy storage. Elsevier, Amsterdam
43. Noh H-J, Youn S, Yoon C, Sun Y-K (2013) Comparison of the structural and electrochemical properties of layered $Li(Ni_xCo_yMn_z)O_2$ ($x=1/3$, 0.5, 0.6, 0.7, 0.8 and 0.85) cathode material for lithium-ion batteries. J Power Sources 233:121–130
44. Schipper F, Erickson E, Erk C, Shin J-Y, Chesneau F, Aurbach D (2017) Review - recent advances and remaining challenges for lithium ion battery cathodes. J Electrochem Soc 164 (1):6220–6228
45. Goriparti S, Miele E, de Angelis F, Di Fabrizio E, Zaccaria RP, Capiglia C (2014) Review on recent progress of nanostructured anode materials for Li-ion batteries. J Power Sources 257:421–443
46. Azuma H, Imoto H, Yamada S, Sekai K (1999) Advanced carbon anode materials for lithium ion cells. J Power Sources 81:1–7
47. Peled E, Golodnitsky D, Ulus A, Yufit V (2004) Effect of carbon substrate on SEI composition and morphology. Electrochim Acta 50:391–395
48. Qi Y, Guo H, Hector L, Timmons A (2010) Threefold increase in the Young's Modulus of graphite negative electrode during Lithium intercalation. J Electrochem Soc 157(5):558–566
49. Liu J, Liu Y, Yang W, Ren Q, Li F, Huang Z (2018) Lithium ion battery separator with high performance and high safety enabled by tri-layered $SiO_2$@PI/m-PE/$SiO_2$@PI nanofiber composite membrane. J Power Sources 396:265–275
50. Fang J, Kelarakis A, Lin Y-W, Kang C-Y, Yang M-H, Cheng C-L, Wang Y, Giannelis E, Tsai L-D (2011) Nanoparticle-coated separators for lithium-ion batteries with advanced electrochemical performance. Phys Chem Chem Phys 13:14457–14461
51. Man C, Jiang P, Wong K-W, Zhao Y, Tang C, Fan M, Lau W-M, Mei J, Li S, Liu H, Dui D (2014) Enhanced wetting properties of a polypropylene separator for a lithium-ion battery by hyper thermal hydrogen induced crosslinking of poly(ethylene oxide). J Mater Chem 2:11980–11986
52. Jeong H-S, Kim D-W, Jeong Y, Lee S-Y (2010) Effect of phase inversion on microporous structure development of $Al_2O_3$/poly(vinylidene fluoridehexafluoropropylene)-based ceramic composite separators for lithium-ion batteries. J Power Sources 195(18):6116–6121
53. Chen W, Liu Y, Ma Y, Yang W (2015) Improved performance of lithium ion battery separator enabled by co-electrospinnig polyimide/poly(vinylidene fluoride-co-hexafluoropropylene) and the incorporation of $TiO_2$ -(2hydroxyethyl methacrylate). J Power Sources 273:1127–1135

54. Wang Y, Wang S, Fang J, Ding L-X, Wang H (2017) A nano-silica modified polyimide nanofiber separator with enhanced thermal and wetting properties for high safety lithium-ion batteries. J Membr Sci 235:248–254
55. Li Q, Chen J, Fan L, Kong X, Lu Y (2016) Progress in electrolytes for rechargeable Li-based batteries and beyond. Green Energy Environ 1:18–42
56. Zhang S (2006) A review on electrolyte additives for lithium-ion batteries. J Power Sources 162:1379–1394
57. Grolleau S, Delaille A, Gualous H, Gyan P, Revel R, Bernard J, Redondo-Inglesias E, Peter J (2014) Calendar aging of commercial graphite/$LiFePO_4$ cell – predicting capacity fade under time dependent storage conditions. J Power Sources 255:450–458
58. Barré A, Deguilhem B, Grolleau S, Gérad M, Suard F, Riu D (2013) A review on lithium-ion battery ageing mechanisms and estimations for automotive applications. J Power Sources 241:680–689
59. Aurback D, Markovsky B, Salitra G, Markevich E, Talyossef Y, Koltypin M (2007) Review on electrode-electrolyte solution interactions, related to cathode materials for Li-ion batteries. J Power Sources 165(2):491–499
60. Zhang J, Terrones M, Park C, Mukherjee R, Monthioux M, Koratkar N, Kim Y, Hurt R, Frackowiak E, Enoki T, Chen Y, Chen Y, Bianco A (2016) Carbon science in 2016: status, challenges and perspectives. Carbon 98(70):708–732
61. Li W, Song B, Manthiram A (2017) High-voltage positive electrode materials for lithium-ion batteries. Chem Soc Rev 46(10):3006–3059
62. Placke T, Kloepsch R, Duehnen S, Winter M (2017) Lithium ion, lithium metal, and alternative rechargeable battery technologies: the odyssey for high energy density. J Solid State Electrochem 21:1939–1964
63. Zheng F, Kotobuki M, Song S, On Lai M, Lu L (2018) Review on solid electrolytes for all-solid-state lithium-ion batteries. J Power Sources 389:198–213
64. Liu X, Zhong L, Huang S, Mao S, Zhu T, Huang J (2012) Size-dependent fracture of silicon nanoparticles during lithiation. Acs Nano 6(2):1522–1531
65. Chan C, Peng H, Liu G, McIlwrath K, Zhang X, Huggins R, Cui Y (2011) High-performance lithium battery anodes using silicon nanowires. In: Materials for sustainable energy: a collection of peer-reviewed research and review articles from nature publishing group. World Scientific, Singapore, pp 187–191
66. Yao Y, Huo K, Hu L, Liu N, Cha J, McDowell M, Chu P, Cui Y (2011) Highly conductive, mechanically robust, and electrochemically inactive tic/c nanofiber scaffold for high-performance silicon anode batteries. ACS Nano 5(10):8346–8351
67. Wen Z, Lu G, Mao S, Kim H, Cui S, Yu K, Huang X, Hurley P, Mao O, Chen J (2013) Silicon nanotube anode for lithium-ion batteries. Electrochem Commun 29:67–70
68. Szcsech J, Jin S (2011) Nanostructured silicon for high capacity lithium battery anodes. Energy Environ Sci 4(1):56–72
69. Burrell A, Cunningham B (2018) Silicon electrolyte interface stabilization (seista): second quarter progress report 2018. SEISta project
70. Wang C, Wu H, Chen Z, McDowell M, Cui Y, Bao Z (2013) Self-healing chemistry enables the stable operation of silicon microparticle anodes for high-energy lithium-ion batteries. Nat Chem 5(12):1042
71. Chen X, Li X, Mei D, Feng J, Hu M, Hu J, Engelhard M, Zheng J, Xu W, Xiao J (2014) Reduction mechanism of fluoroethylene carbonate for stable solid electrolyte interphase film on silicon anode. ChemSusChem 7(2):549–554
72. Zong L, Jin Y, Liu C, Zhu B, Hu X, Lu Z, Zhu J (2016) Precise perforation and scalable production of si particles from low-grade sources for high-performance lithium ion battery anodes. Nano Lett 16(11):7210–7215
73. Dash R, Pannala S (2016) The potential of silicon anode based lithium ion batteries. Mater Today 9(19):483–484

74. Chae S, Ko M, Park S, Kim N, Ma J, Cho J (2016) Micron-sized fe-cu-si ternary composite anodes for high energy li-ion batteries. Energy Environ Sci 9(4):1251–1257
75. Li J, Dahn J (2007) An in situ x-ray diffraction study of the reaction of li with crystalline si. J Electrochem Soc 154(3):A156–A161
76. Yoon T, Nguyen C, Seo D, Lucht B (2015) Capacity fading mechanisms of silicon nanoparticle negative electrodes for lithium ion batteries. J Electrochem Soc 162(12):A2325–A2330
77. Philippe B, Dedryvére R, Gorgoi M, Rensmo H, Gonbeau D, Edstroöm K (2013) Role of the lipf6 salt for the long-term stability of silicon electrodes in li-ion batteries - a photoelectron spectroscopy study. Chem Mater 25(3):394–404
78. Jin Y, Zhu B, Lu Z, Liu N, Zhu J (2017) Challenges and recent progress in the development of si anodes for lithium-ion battery. Adv Energy Mater 7(23):1700715
79. Kim H, Choi S, Lee S, Seo M, Lee J, Deniz E, Lee Y, Kim E, Choi J (2015) Controlled prelithiation of silicon monoxide for high performance lithium-ion rechargeable full cells. Nano Lett 16(1):282–288
80. Forney M, Ganter M, Staub J, Ridgley R, Landi B (2013) Delithiation of silicon-carbon nanotube anodes for lithium ion batteries by stabilized lithium metal powder (slmp). Nano Lett 13(9):4158–4163
81. Zhao J, Zhenda L, Liu N, Lee H-W, McDowell M, Cui Y (2014) Dry-air-stable lithium silicide-lithium oxide core-shell nanoparticles as high-capacity prelithiation reagents. Nat Commun 5:5088
82. Li X, Colclasure A, Finegan D, Ren D, Shi Y, Feng X, Cao L, Yang Y, Smith K (2019) Degradation mechanisms of high capacity 18650 cells containing Si-graphite anode and nickel-rich NMC cathode. Electrochim Acta 297:1109–1120
83. Manthiram A, Song B, Li W (2017) A perspective on nickel-rich layered oxide cathodes for lithium-ion batteries. Energy Storage Mater 6:125–139
84. Song H, Heui S, Ahn J, Oh S, Yim T (2019) Artificial cathode-electrolyte interphases on nickel-rich cathode materials modified by silyl functional group. J Power Sources 416:1–8
85. Yu X, Zheng J, Wang C, Gu M (2018) Designing principle for Ni-rich cathode materials with high energy density for practical applications. Nano Energy 49:434–452
86. Zhao W, Zheng G, Lin M, Zhao W, Li D, Guan X, Ji Y, Ortiz G, Yang Y (2018) Toward a stable solid-electrolyte-interfaces on nickelrich cathodes: LiPO2F2 salt-type additive and its working mechanism for LiNi0.5Mn0.25Co0.25O2 cathodes. J Power Sources 380:149–157
87. Bak S-M, Hu E, Zhou Y, Yu X, Senanayake S, Cho S-J, Kim K-B, Chung K, Yang X-Q, Nam K-W (2014) Structural changes and thermal stability of charged LixMnyCozO2 cathode materials studied by combined in situ time-resolved XRD and mass spectroscopy. ACS Appl Mater Interfaces 6(24):22594–22601

# Part IV

# Micro-optical Sensors in Energy Systems

# Thermal Management of Lithium-Ion Batteries

12

Lena Spitthoff, Eilif S. Øyre, Harald I. Muri, Markus Wahl, Astrid F. Gunnarshaug, Bruno G. Pollet, Jacob J. Lamb, and Odne S. Burheim

## Introduction

Li-ion batteries have become the cornerstone of electrical energy storage in recent decades, resulting in a significant transition to hybrid and fully electric cars. Furthermore, the energy density of batteries, in general, has developed significantly from around 30 Wh $kg^{-1}$ for lead-based batteries, up to over 200 Wh $kg^{-1}$ for Li-ion batteries [1]. Because of these significant increases in specific energy (as well as reductions in cost and improvements in durability), Li-ion-based batteries have

L. Spitthoff · E. S. Øyre
Department of Energy and Process Engineering, ENERSENSE, Norwegian University of Science and Technology, Trondheim, Norway
e-mail: lena.spitthoff@ntnu.no; eilifso@stud.ntnu.no

H. I. Muri · M. Wahl
Department of Electronic Systems, ENERSENSE, Norwegian University of Science and Technology, Trondheim, Norway
e-mail: harald.muri@ntnu.no; markus.s.wahl@ntnu.no

A. F. Gunnarshaug
Department of Chemistry, Norwegian University of Science and Technology, Trondheim, Norway
e-mail: astrid.f.gunnarshaug@ntnu.no

B. G. Pollet · O. S. Burheim
Department of Energy and Process Engineering, ENERSENSE and NTNU Team Hydrogen, Norwegian University of Science and Technology, Trondheim, Norway
e-mail: bruno.g.pollet@ntnu.no; odne.s.burheim@ntnu.no

J. J. Lamb (✉)
Department of Electronic Systems, ENERSENSE, Norwegian University of Science and Technology, Trondheim, Norway

Department of Energy and Process Engineering, ENERSENSE, Norwegian University of Science and Technology, Trondheim, Norway
e-mail: jacob.j.lamb@ntnu.no



J. J. Lamb, B. G. Pollet (eds.), *Micro-Optics and Energy*,
https://doi.org/10.1007/978-3-030-43676-6_12

already been implemented into small transport vehicles. Presently Li-ion batteries are being implemented into large-scale hybrid and electric vehicles [2], such as electric buses, hybrid electric buses and hybrid-powered ships [3], as bigger cells have become cost-effective. Because bigger cars use electricity, there is a need for bigger battery packs that can withstand more severe usage. To realise the full potential of Li-ion batteries, thermal management of their internal and external environments is required. To achieve this, small sensors (e.g. 10 μm thick), stable and inert are required. In this chapter, thermal management with regard to the structure of Li-ion batteries will be discussed, and how micro-optical sensors may facilitate improvements of the thermal management.

## System Description

Rechargeable Li-ion batteries (termed secondary Li-ion batteries) come in a variety of different shapes and sizes, where the capacity typically defines the size in terms of Ah, Coulombs or kWh. An early adopted Li-ion battery standard was the 18,650 cell, which was a cylindrical battery with a capacity of 1.5–3.0 Ah. In such cells the battery electrodes and separator are rolled up (Fig. 12.1, right); however, modern Li-ion batteries are much larger and generally shaped like rectangular prisms. Inside these, there can be multiple cells. These cells are composed of the battery electrodes and separators and are held together in flat layers (Fig. 12.1, left). The layer stack is either sealed in a pouch or a prismatic container, forming pouch cell batteries and prismatic batteries, respectively. The prismatic and pouch cell batteries have larger capacity whether they are for cellular phones, computers, cars or other heavy-duty systems. The capacity can be from 4 to 20 Ah, and (in some specially designed batteries), up to 100 Ah.

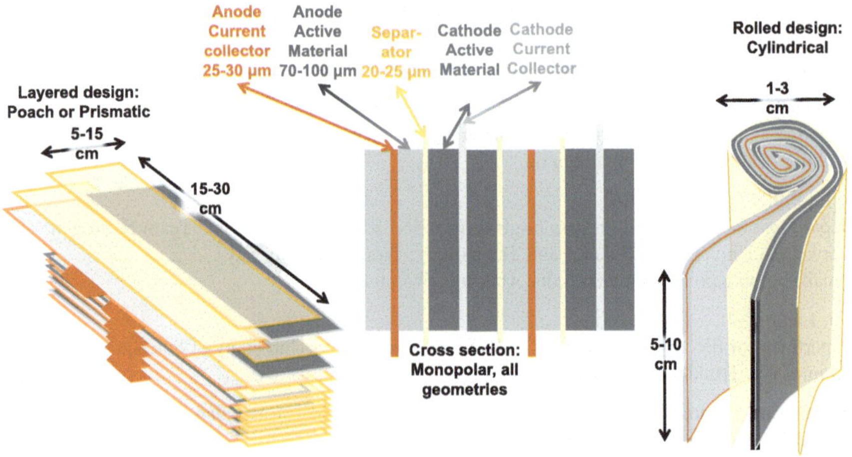


**Fig. 12.1** Overview of Li-ion battery components and assembly configurations

The secondary Li-ion battery has two types of electrodes (an anode and a cathode), which are defined by the discharge process. The anode is defined as the electrode where the oxidation process happens during discharge, which in the case of a Li-ion battery is a carbon-based intercalation anode. This means that the lithium ions, provided by electrolyte salt (lithium hexafluorophosphate; $LiPF_6$), are stored by intercalating in the carbon-based anode during charging. During discharge, the Li is oxidised and liberated from the anode. The liberated Li-ions migrate through the electrolyte-separator region towards the cathode. At the cathode, some of the metals in the mixed metal oxides are reduced and change their oxidation number. The earliest commercial secondary Li-ion battery consisted of a mixed metal cathode of $LiCoO_2$ when fully discharged, and $Li_{0.45}CoO_2$ when fully charged [4]. It is cobalt that changes oxidation number in the cathode reduction reaction (from 3.55 to 3).

The lithium cobalt oxide battery is generally termed LCO. When cobalt is replaced with nickel to make a mixture of nickel, manganese and cobalt it is termed NMC, when manganese is replaced by aluminium, it is named NCA, and when the mixture is replaced by iron phosphate ($FePO_4$), the battery is termed LFP.

The design of the Li-ion battery is based on diffusing the active electrode material in a solvent and then smearing this onto a metal film (e.g. by using a doctor blade [5]). This is done on both sides of the metal film. Next the electrodes are layered with a separator in between (Fig. 12.1, middle). For the anode, the active electrode material is carbon based, and the substrate metal film is copper. The current collector (or copper substrate) is typically 30 μm or less, and the active anode carbon layer is 70–100 μm thick [6]. For the cathode, the active layer is typically a mixed metal oxide with a substrate metal film of aluminium [1]. The aluminium film is also 30 μm or less, and the active material (cathode) is between 90 and 130 μm. The two active electrodes are kept apart by a separator, which is a non-conductive porous layer soaked in electrolyte, typically sized 20–25 μm [7, 8].

This design for assembling electrodes and separators is conventional for all Li-ion batteries. To undertake in situ and *in-operando* measurements of temperature inside a Li-ion battery, the sensor must be as small as possible, preferably in a range smaller than the separator (i.e. 20–25 μm). Modifying optical sensors to fit this type of geometric constraint opens for measurements that have negligible interference with the normal functioning of the Li-ion battery.

## Importance of Thermal Management

In both the internal and external environment of the battery, thermal management becomes much more significant. The increasing use of Li-ion batteries is not only owing to their trait of zero-emissions during operation, and their low carbon footprint [9], a more cost-effective implementation has been shown as possible [3]. Despite this, the battery's specific energy remains a limiting factor when compared to fossil fuel vehicles [10]. Fast charging requires a good understanding of heat production and transfer within the battery. Effects such as capacity drop, power drop and self-discharge in Li-ion batteries are well documented in the literature [11]. The impact of temperature on various ageing processes, in particular, is well reported [12–15].

The performance and life of a battery depend heavily on the design [16, 17]. Moreover, the design will affect current availability, temperature production, state of charge (SOC) and voltage delivery. The local temperatures and local degradation (ageing) of the battery will suffer as a result [18]. The right determination and distribution of heat sources together with the thermal conductivities of the various parts will enable inner temperature profiles to be modelled. For example, the Seebeck coefficient can be related to the local reversible Peltier heat at the electrode interface inside a Li-ion battery. The Seebeck coefficient can be measured using a symmetric thermoelectric cell, improving computational modelling of Li-ion cells [19]. This can assist cooling schemes outside the battery or battery pack. Also, improved thermal management will enhance local ageing process knowledge and may contribute to stronger battery designs and increased lifetimes. Knowing inner temperature profiles of Li-ion batteries at elevated current densities also allows for predicting and avoiding thermal runaway situations, which is the primary safety concern for Li-ion batteries during elevated current density operations. Furthermore, during ageing, internal resistance has been observed to rise, and this will also affect the heat output within the battery [6, 20, 21].

All thermal conductivities of components are essential parameters for temperature profile modelling of Li-ion batteries. The thermal conductivity of the materials is not necessarily isotropic. The words 'in-plane' and 'cross-plane' are usually used. For example, in a thin electrode, we distinguish between the perpendicular (cross-plane) orientation and the parallel (in-plane) orientation. There are accounts of Li-ion battery components' heat conductivity, but they are not studied thoroughly [8, 11]. More specifically, there are not so many accounts on the thermal conductivity of separators. There are many different values given as the thermal conductivity of separators ranging from 0.33 to 1.29 $WK^{-1}m^{-1}$ [22–25]. Some of these characteristics are claimed, and the laboratory determination of a separator's cross-plane heat conductivity is hardly reported. Due to the predominance of heat production in the separator-electrolyte region and to some extent at high currents within the solid electrolyte interface region [8], knowledge of the electrolyte-soaked separator's thermal conductivity is critical.

## Calculating the Internal Heat Production

Within a secondary Li-ion battery, there are three sources of heat:

1. Ohmic production of heat
2. Entropy change of the electrode reactions' production or consumption of heat (reversible)
3. The resistance of the electrochemical reaction, leading to heat production

The total heat generation of a Li-ion battery during charging and discharging can be encompassed by:

$$q_{\text{ch}} = \left(\frac{T \times \Delta S}{F} j + r_{\omega} j^2 + \eta j\right) \tag{12.1}$$

$$q_{\text{disch}} = \left(-\frac{T \times \Delta S}{F} j + r_{\omega} j^2 + \eta j\right) \tag{12.2}$$

Here, $j$ is the current density, $r_{\omega}$ is the ohmic resistance, $T$ is the temperature, $\Delta S$ is the entropy change of the whole cell, $F$ is the Faraday constant and $\eta$ is the overpotential [7].

## Thermal Conductivity Measurement

The total thermal resistance is the sum of the components' thermal resistance and its contact resistance to a measuring sensor. By sampling the same component in various thicknesses, the total thermal resistance can be plotted as a function of the component thickness (this can be achieved by stacking the component multiple times). The thermal conductivity can then be obtained by calculating the inverse of the slope from this plot. Concerning the total thermal resistance of electrodes, the thermal resistance consisted of the contact resistance, the thermal resistance of the active material and the current collector thermal resistance. The total thermal resistance can be calculated by measuring the heat flux through the sample and the temperature drop over the sample [6]:

$$R_{\text{sample}} = \frac{T_4 - T_5}{q_{\text{sample}}} = 2R_{\text{contact}} + \frac{\delta_{\text{act.M.}}}{k_{\text{act.M.}}} + \frac{\delta_{\text{foil}}}{k_{\text{foil}}} \tag{12.3}$$

Here, $R_{\text{sample}}$ is the total thermal resistance, $R_{\text{contact}}$ is the contact resistance, $\frac{\delta_{\text{act.M.}}}{k_{\text{act.M.}}}$ is the thermal resistance of the active material in an electrode and $\frac{\delta_{\text{foil}}}{k_{\text{foil}}}$ is the thermal resistance of the current collector (where $\delta$ is the thickness and $\delta$ is the thermal conductivity). It has been suggested that additional resistances from contact in a Li-ion battery stack can be neglected [26].

# Determining Parameters Experimentally with Micro-optic Sensors

## Temperature Characterisation Requirements

Temperature measurements within the Li-ion battery are required to obtain a three-dimensional distribution of heat within the battery [6–8]. The sensor should not change the electrochemical and physical environment (leading to changes in the operation and thermal conductivity) within the Li-ion battery. To best achieve this,

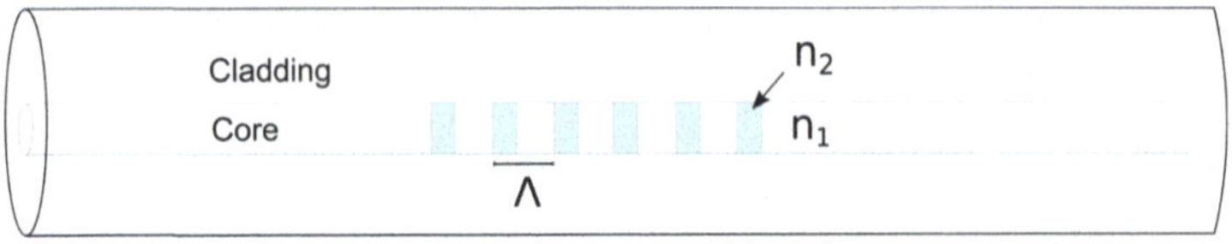


**Fig. 12.2** Schematic of an optical fibre with inscribed fibre Bragg grating

the sensor should be inert, of small size and not electrically conductive to avoid short-circuiting of the separator.

## Fibre Bragg Grating Sensor

Creating a grating corrugation in the refraction index ($n_1$ and $n_2$ on Fig. 12.2) of the core of a single-mode optical fibre causes a successive, coherent scattering of a narrow band of the incident field [27]. The strongest reflection occurs at the Bragg wavelength $\lambda_B$, given by the equation

$$\lambda_B = 2\eta_{eff}\Lambda \tag{12.4}$$

As a result, the reflected wavelength depends on the periodicity of the grating, $\Lambda$, and the modal index, $\eta_{eff}$, which in turn both depend on temperature $T$ and strain $\epsilon$. The change in $\lambda_B$ as a result of temperature and strain can be written as:

$$\Delta\lambda_B = 2\left(\Lambda\frac{d\eta_{eff}}{dT} + \eta_{eff}\frac{d\Lambda}{dT}\right)\Delta T + 2\left(\Lambda\frac{d\eta_{eff}}{d\epsilon} + \eta_{eff}\frac{d\Lambda}{d\epsilon}\right)\Delta\epsilon \tag{12.5}$$

The temperature dependence is linear and without hysteresis.

## Temperature Gradient Ratio

Applying the thermal conductivity and resistance of cell parts together with the pouch cell thickness 2.3 mm yields the ratio of the temperature gradient across the electrodes ($\Delta T$), over the temperature gradient across external thermocouples $\Delta T_{tot}$ [7, 19]:

$$\frac{\Delta T}{\Delta T_{tot}} = 0.92 \tag{12.6}$$

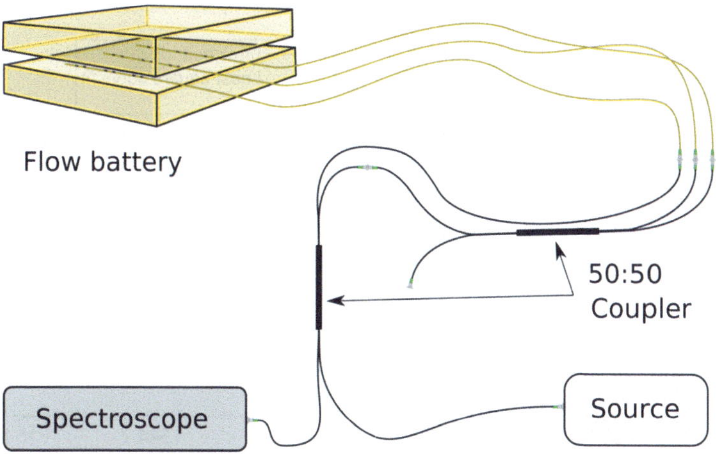


**Fig. 12.3** Schematic of the experimental setup

## Experimental Setup

Two 125 μm in diameter FBG sensors were implemented in the pouch cell between the aluminium pouch laminate and the electrodes, before sealing two of three sides. The electrolyte was added to the separator in a moisture-free glovebox before sealing the final side in a vacuum. The pouch cell was assembled between two copper plates at which the temperature was controlled by flowing water from external water reservoirs through aluminium plates in contact with the copper, see Fig. 12.3. Thermocouples were installed such that they were separated by the 2.3 mm pouch cell and the two 2 mm copper plates.

An 85 nm bandwidth signal centred around 1550 nm was sent from a source through the fibre. The reflected signal passed back to a spectroscope, and the Bragg wavelength $\lambda_B$ was found using a Gaussian fit of the reflected spectrum.

## Calibration

The sensors were calibrated by thermostat using multiple heat baths equally on different temperatures with a precision ±0.1 K. The average temperature value from the thermocouples were used as reference with corresponding uncertainty ±0.1. Figure 12.4 shows the temperature response of $\lambda_B$ for FBG sensor.

Linear regression with root mean square error of ±0.12 K was calculated using the least square method. The temperature sensitivity of the sensor is given by the slope of the regression, in this case 31.671 ± 0.006 K per nm. The error bounds of the curve are calculated based on the uncertainty in the regression coefficients. Within the calibration interval, this error equals 0.06 K. Temperature evolution at

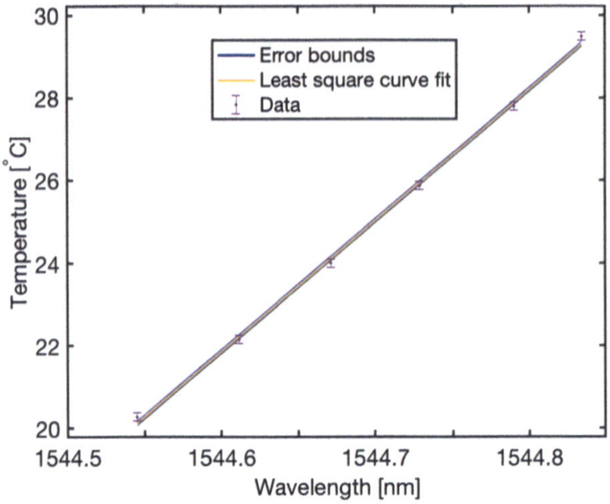


**Fig. 12.4** Temperature response of $\lambda_B$ for FBG sensor. The error bars are based on the uncertainty in the curve coefficients, being ±0.06 K within this calibration interval

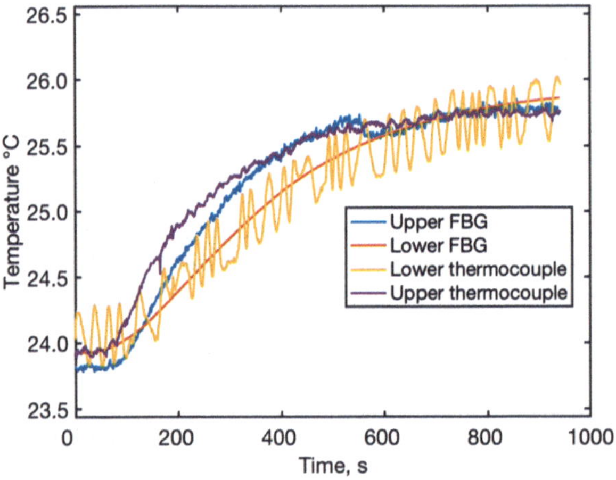


**Fig. 12.5** Temperature evolution of the sensors after 2 K external heat bath temperature increase

the thermocouples, and two FBG sensors, one above the electrodes and one below, after a 2 K increase in heat bath temperature is shown in Fig. 12.5.

## Implementation of Sensors into a Li-Ion Pouch

The lower FBG sensor in this cell had an uncertainty of ±0.02 K, seven times lower than the upper FBG sensor with ±0.14 K. This is because the latter had a lower intensity in the reflected spectrum. The sensors follow the temperature at the closest thermocouple with a small delay and present a way of measuring the temperature inside the cell while giving far more stable and precise result than the thermocouples.

One of the fibres in the pouch cell with two sensors snapped such that the gradient across the electrodes had to be found by measuring a single side before flipping the

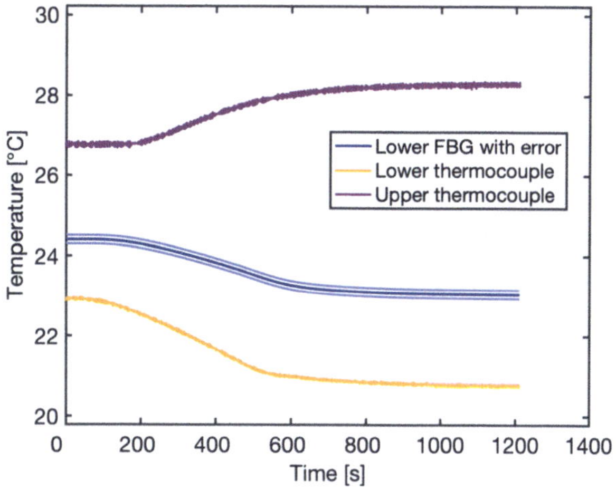


**Fig. 12.6** Temperature evolution of the sensors after increasing $\Delta T_{ext}$ with 5 K

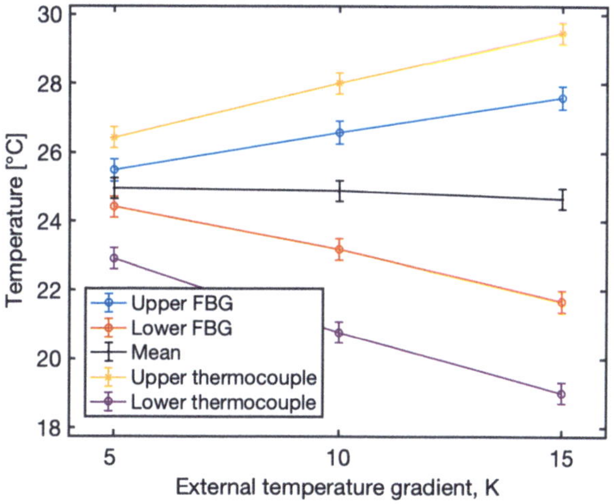


**Fig. 12.7** Temperature stability at different $\Delta T_{ext}$

pouch cell and repeating the measurement. Figure 12.6 shows how the thermocouples and FBG sensor responds after doubling the externally applied temperature gradient. The shaded area illustrates the error bounds of the FBG sensor.

The resulting stable values during three different external gradients averaging 25 K are shown with uncertainties in Fig. 12.7. The mean corresponds to the average value of the upper and lower FBG sensor. The calculated gradients and their uncertainties are listed in Table 12.1.

The ratios are lower than the theoretical ratio, which may be caused by not including housing contact resistance [19]. Another source of error is the formation of air pockets inside the cell. Also, different ratios for different external temperatures may indicate that the temperature gradient inside the cell was not fully established at the time of measurement, or a systematic error in one of the thermocouples.

**Table 12.1** Calculated gradients and their uncertainties

| $\Delta T_{ext}$ | $\Delta T_e$ | $\Delta T_{tot}$ | $\Delta T_e/\Delta T_{tot}$ |
|---|---|---|---|
| 5 | 1.07 ± 0.45 | 3.53 | 0.30 ± 0.13 |
| 10 | 3.40 ± 0.45 | 7.23 | 0.47 ± 0.07 |
| 15 | 5.89 ± 0.45 | 10.46 | 0.56 ± 0.05 |

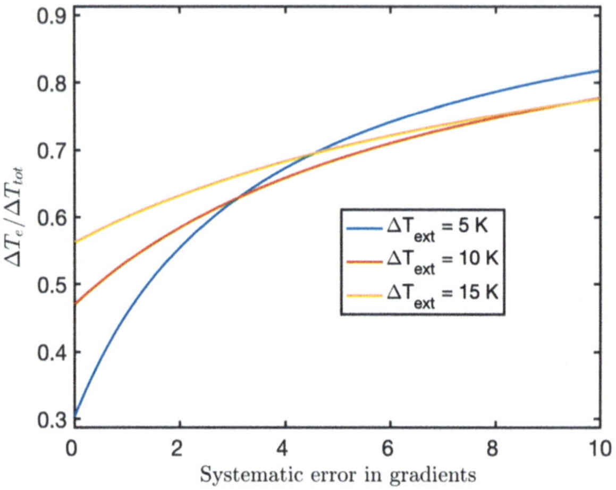


**Fig. 12.8** The graphs show what the accurate ratios would be by removing a corresponding systematic error

Figure 12.8 plots the change in $\Delta T_e/\Delta T_{tot}$ as a function of systematic error. A small systematic error could explain some of the instabilities in the three ratios but does not explain why the ratios are generally low.

## Conclusion

The fibres are the most vulnerable during the sealing of the pouch, and a protective coating, which is inert to the reactions within the cell, is recommended around the seal. The calibrated FBG sensors had a measurement precision ranging from 0.14 to 0.02 K, showing great potential for real-time temperature measurement. The precision and accuracy of the sensors were restricted by the reference thermocouples.

**Acknowledgements** The authors are grateful to the ENERSENSE programme and NTNU Team Hydrogen at the Norwegian University of Science and Technology (NTNU) for supporting and helping on this book project.

## References

1. Burheim OS (2017) Engineering energy storage. Academic Press, Oxford
2. Pollet BG, Staffell I, Shang JL (2012) Current status of hybrid, battery and fuel cell electric vehicles: from electrochemistry to market prospects. Electrochim Acta 84:235–249

3. Vartdal BJ, Chryssakis C (2011) Potential benefits of hybrid powertrain systems for various ship types. In: International scientific conference on hybrid and electric vehcicles, Paris, France, pp 1–12
4. Goodenough JB, Park K-S (2013) The Li-ion rechargeable battery: a perspective. J Am Chem Soc 135(4):1167–1176
5. Wood III DL, Li J, Daniel C (2015) Prospects for reducing the processing cost of lithium ion batteries. J Power Sources 275:234–242
6. Richter F, Vie PJS, Kjelstrup S, Burheim OS (2017) Measurements of ageing and thermal conductivity in a secondary NMC-hard carbon Li-ion battery and the impact on internal temperature profiles. Electrochim Acta 250:228–237
7. Richter F, Kjelstrup S, Vie PJS, Burheim OS (2017) Thermal conductivity and internal temperature profiles of Li-ion secondary batteries. J Power Sources 359:592–600
8. Burheim OS, Onsrud MA, Pharoah JG, Vullum-Bruer F, Vie PJS (2014) Thermal conductivity, heat sources and temperature profiles of Li-ion batteries. ECS Trans 58(48):145–171
9. Saevarsdottir G, Tao P, Stefansson H, Harvey W (2014) Potential use of geothermal energy sources for the production of lithium-ion batteries. Renew Energy 61:17–22
10. Mayer T, Kreyenberg D, Wind J, Braun F (2012) Feasibility study of 2020 target costs for PEM fuel cells and lithium-ion batteries: a two-factor experience curve approach. Int J Hydrog Energy 37(19):14463–14474
11. Bandhauer TM, Garimella S, Fuller TF (2011) A critical review of thermal issues in lithium-ion batteries. J Electrochem Soc 158(3):R1–R25
12. Broussely M, Biensan P, Bonhomme F, Blanchard P, Herreyre S, Nechev K et al (2005) Main aging mechanisms in Li ion batteries. J Power Sources 146(1–2):90–96
13. Vetter J, Novák P, Wagner MR, Veit C, Möller K-C, Besenhard JO et al (2005) Ageing mechanisms in lithium-ion batteries. J Power Sources 147(1–2):269–281
14. Leng F, Tan CM, Pecht M (2015) Effect of temperature on the aging rate of Li ion battery operating above room temperature. Sci Rep 5:12967
15. Waldmann T, Wilka M, Kasper M, Fleischhammer M, Wohlfahrt-Mehrens M (2014) Temperature dependent ageing mechanisms in Lithium-ion batteries–a post-mortem study. J Power Sources 262:129–135
16. Zhang G, Shaffer CE, Wang C-Y, Rahn CD (2013) Effects of non-uniform current distribution on energy density of Li-ion cells. J Electrochem Soc 160(11):A2299–A2305
17. Nanda J, Remillard J, O'Neill A, Bernardi D, Ro T, Nietering KE et al (2011) Local state-of-charge mapping of Lithium-ion battery electrodes. Adv Funct Mater 21(17):3282–3290
18. Robinson JB, Darr JA, Eastwood DS, Hinds G, Lee PD, Shearing PR et al (2014) Non-uniform temperature distribution in Li-ion batteries during discharge–a combined thermal imaging, X-ray micro-tomography and electrochemical impedance approach. J Power Sources 252:51–57
19. Richter F, Gunnarshaug A, Burheim OS, Vie PJS, Kjelstrup S (2017) Single electrode entropy change for $LiCoO_2$ electrodes. ECS Trans 80(10):219–238
20. Eddahech A, Briat O, Vinassa J-M (2015) Performance comparison of four lithium–ion battery technologies under calendar aging. Energy 84:542–550
21. Tanim TR, Rahn CD (2015) Aging formula for lithium ion batteries with solid electrolyte interphase layer growth. J Power Sources 294:239–247
22. Chen SC, Wan CC, Wang YY (2005) Thermal analysis of lithium-ion batteries. J Power Sources 140(1):111–124
23. Taheri P, Bahrami M (2012) Temperature rise in prismatic polymer lithium-ion batteries: An analytic approach. SAE Int J Passeng Cars-Electronic Electr Syst 5(1):164–176. https://doi.org/10.4271/2012-01-0334
24. Kim US, Shin CB, Kim C-S (2009) Modeling for the scale-up of a lithium-ion polymer battery. J Power Sources 189(1):841–846

25. Wu W, Xiao X, Huang X (2011) Modeling heat generation in a lithium ion battery. In: ASME 2011 5th international conference on energy sustainability. American Society of Mechanical Engineers, Washington, DC, pp 1513–1522
26. Burheim O, Vie PJS, Pharoah JG, Kjelstrup S (2010) Ex situ measurements of through-plane thermal conductivities in a polymer electrolyte fuel cell. J Power Sources 195(1):249–256
27. Hill KO, Meltz G (1997) Fiber Bragg grating technology fundamentals and overview. J Light Technol 15(8):1263–1276

# Reverse Electrodialysis Cells

# 13

Kjersti W. Krakhella, Markus Wahl, Eilif S. Øyre, Jacob J. Lamb, and Odne S. Burheim

## Introduction

In a closed system, reverse electrodialysis (RED) and electrodialysis (ED) can be coupled into an energy storage system where excess power can be placed in a salinity gradient between two solutions. The device is powered by applying an external power source to force ions from a dilute solution to a concentrated solution through alternating ion exchange membranes. The current is discharged by allowing ions to move from the concentrated solution through the same membranes to dilute solutions, supplying electrical energy at the electrodes. Most literature claims have

K. W. Krakhella
Department of Materials Science and Engineering, ENERSENSE, Norwegian University of Science and Technology, Trondheim, Norway

M. Wahl
Department of Electronic Systems, ENERSENSE, Norwegian University of Science and Technology, Trondheim, Norway
e-mail: markus.s.wahl@ntnu.no

E. S. Øyre
Department of Energy and Process Engineering, ENERSENSE, Norwegian University of Science and Technology, Trondheim, Norway
e-mail: eilifso@stud.ntnu.no

J. J. Lamb (✉)
Department of Electronic Systems, ENERSENSE, Norwegian University of Science and Technology, Trondheim, Norway

Department of Energy and Process Engineering, ENERSENSE, Norwegian University of Science and Technology, Trondheim, Norway
e-mail: jacob.j.lamb@ntnu.no

O. S. Burheim
Department of Energy and Process Engineering, ENERSENSE and NTNU Team Hydrogen, Norwegian University of Science and Technology, Trondheim, Norway
e-mail: odne.s.burheim@ntnu.no



J. J. Lamb, B. G. Pollet (eds.), *Micro-Optics and Energy*,
https://doi.org/10.1007/978-3-030-43676-6_13

been developed using open real-world concentration gradients (seawater and freshwater), whereas a closed system may have a higher salt concentration (NaCl up to 6 M). A closed system also benefits from the absence of membrane fouling, and it is simpler to monitor and maintain the temperature. In this chapter, the closed reverse electrodialysis system is discussed.

## Salinity Gradient Energy Storage

A salinity gradient-based energy storage system incorporates a desalination technology with a mixing technology. The device is loaded with internal power that increases the gap in concentration between two alternatives, and energy is conserved as a chemical potential difference between two concentrated solutions. By blending the two alternatives, turning the chemical potential into electricity and reducing the distinction in density, the chemical energy is released as electricity. Electrodialytic energy storage systems (EESSs; including RED and ED systems) are still at the primary research phase, with their development still in their infancy. Their potential power output is highly dependent on the concentration difference between the two solutions in the EESS, with large concentration differences having a higher stored chemical energy [1, 2].

## Electrodialytic Energy Storage System Principles

Modelling of the EESS system is essential for optimisation and development of the system. During discharge, the chloride ions pass through the anion interchange membrane (AEM) from the concentrated solution to the dilute solution, while the sodium ions pass through the cation exchange membrane (CEM) from the concentrated water to the dilute. A redox or rinse fluid is circulated on each electrode. Both $FeCl_2$ and $FeCl_3$ are used as the redox solution. Figure 13.1 shows an EESS during discharge.

The chloride ions enter the redox fluid at the anode during discharge, while chloride ions leave the redox fluid at the cathode. This is shown as:

$$\text{Anode}: Fe^{2+} \rightarrow Fe^{3+} + e^{-} \tag{13.1}$$

$$\text{Cathode}: Fe^{3+} + e^{-} \rightarrow Fe^{2+} \tag{13.2}$$

During discharge, the electrons move from the anode to the cathode. When charging the EESS, electrons are forced from the anode to the cathode, also reversing the reactions provided in Eqs. (13.1 and 13.2).

The open-circuit voltage (OCV) is derived from the Nernst equation:

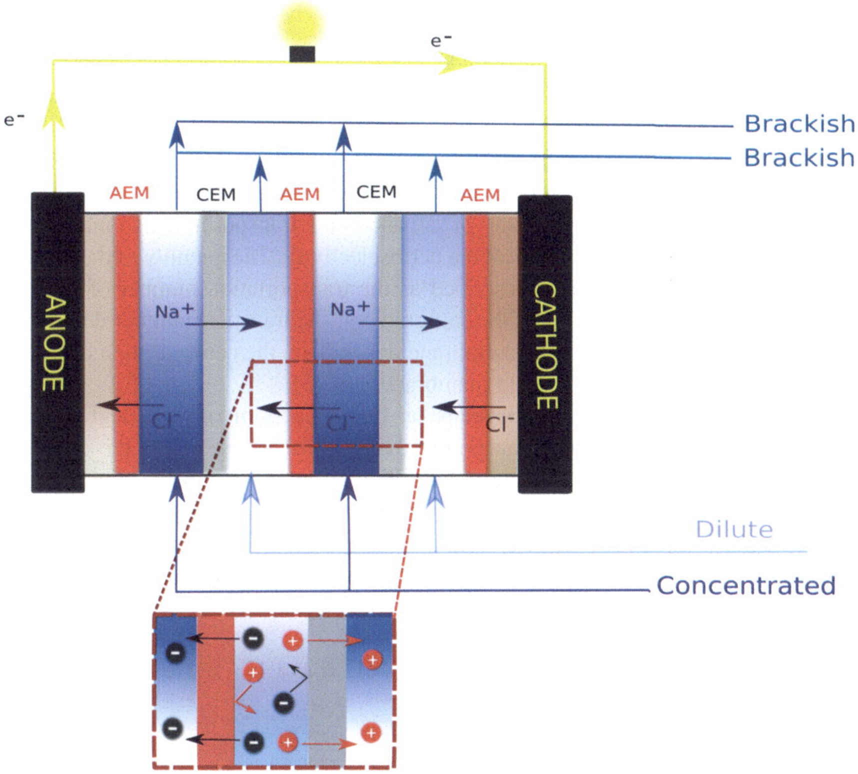


**Fig. 13.1** The discharging process of an EESS

$$E_{\mathrm{OCV}} = 2\overline{\alpha}\frac{RT}{zF}\ln\left(\frac{c_c\gamma_c}{c_d\gamma_d}\right) \tag{13.3}$$

Here, $R$ is the ideal gas constant, $T$ is the temperature (Kelvin), $F$ is the constant of Faraday, $z$ is the number of electrons transmitted, and $\alpha$ is the average CEM and AEM permselectivity. The membrane permselectivity reflects the ability to discriminate between two electrons. It depends heavily on the density of the solution and the amount of salt and water transportation. The concentrated and diluted solution's molarity and activity coefficients are $c_c$ and $c_d$, and $\gamma_c$ and $\gamma_d$, respectively. By increasing the activity difference between the two solutions used in EESS, the OCV and power density will increase. More specifically, it is ideal to have $c_c$ as high as possible, while $c_d$ is kept close to 1 M. Temperature also affects the activity significantly in a linear relationship.

The permselectivity for an ion exchange membrane can be theoretically calculated by:

$$\alpha = t_s - t_w \times M_W \times \frac{b_1 - b_2}{\ln\left(\frac{(b_1 \times \gamma_1)}{(b_2 \times \gamma_2)}\right)} \quad (\text{when } b_1 \neq b_2) \tag{13.4}$$

$$\alpha = t_s - t_w \times M_W \times b_1 \quad (\text{when } b_1 = b_2) \tag{13.5}$$

Here, $t_s$ and $t_w$ are the number for salt and water transportation, $M_W$ is the molar water weight (0.018 kg/mol) and $b_1$ and $b_2$ are the molar solutions on each side of the membrane. Due to the steady amount of salt and water transportation, an increase in concentration difference would boost the permselectivity. The quantity of the current conducted by the counterion is described as the transportation number of salts. If more co-ions are transported with the salt gradient, the transport number decreases, and the quantity of co-ions in the membrane depends on the concentration of salt in excess and the quantity of the fixed membrane ions [3].

The water transport number depends on the impacts of osmotic and electro-osmotic effects [4]. The water transported by osmosis diffuses from the dilute to the concentrated solution, reducing permselectivity. At the same time, water transported by electroosmosis is transported around the counterions as a hydration shell (hydration number), moving from the concentrated solution to the diluted solution, increasing permselectivity. With more considerable concentration differences, the osmotic impact will become more significant, while the number of water molecules per ion will be lowered [5].

Concerning the total unit cell voltage, Eqs. (13.6 and 13.7) give the voltage for charging and discharging, respectively:

$$E^{\text{charging}} = E_{\text{OCV}} + r\Omega \times i^{\text{charging}} \tag{13.6}$$

$$E^{\text{discharging}} = E_{\text{OCV}} + r\Omega \times i^{\text{discharging}} \tag{13.7}$$

Here, $i$ is the current density measured per area of membrane and $r\Omega$ is the ohmic resistance per $m^2$ of a cell unit. The ohmic resistance is the sum of the resistance of one AEM, one CEM, a dilute and a concentrated compartment. Although the electrode resistance can be assumed negligible in large-scale stacks [6], it is necessary to include it when calculating the total unit cell voltage in small-scale laboratory stacks [7].

The power density ($W/m^2$) for charging and discharging can be calculated from the electrode voltage and the current density:

$$P^{\text{charging}} = E_{\text{OCV}} i^{\text{charging}} + r\Omega \times \left(i^{\text{charging}}\right)^2 \tag{13.8}$$

$$P^{\text{discharging}} = E_{\text{OCV}} i^{\text{discharging}} - r\Omega \times \left(i^{\text{discharging}}\right)^2 \tag{13.9}$$

The current density while charging is approximately half the current density while discharging; therefore, the discharging current density at peak power can be calculated as:

$$i^{\text{discharging}} = \frac{E_{\text{OCV}}}{2r\Omega} \tag{13.10}$$

Taking this into account, the power density equations (Eqs. 13.8 and 13.9) can be rewritten as:

$$P^{\text{charging}} = \frac{E_{\text{OCV}}^2}{4r\Omega} \tag{13.11}$$

$$P_{\text{max}}^{\text{discharging}} = \frac{5E_{\text{OCV}}^2}{16r\Omega} \tag{13.12}$$

Whereas the total available power density at open-circuit voltage is:

$$P_{\text{available}} = E_{\text{OCV}} i \tag{13.13}$$

The EESS discharging and charging efficiencies are reversible, can be calculated by Eqs. (13.14 and 13.15):

$$\eta_{\text{discharging}} = \frac{P_{\text{max}}^{\text{discharging}}}{P_{\text{available}}} \tag{13.14}$$

$$\eta_{\text{charging}} = \frac{P_{\text{available}}}{P^{\text{charging}}} \tag{13.15}$$

Any extra power usage for the system (e.g. power supply to solution pumps and polarisation losses), must be subtracted from the $P_{\text{max}}^{\text{discharging}}$ or $P^{\text{charging}}$ to acquire an accurate efficiency value.

## Determining Parameters Experimentally with Micro-optic Sensors

A grid of sensors was desired to determine the temperature within an EESS. EESSs, in particular, a concentration battery through a coupling of RED and ED, are flow batteries with significant potential to replace ordinary electrochemical batteries for large storage systems. Despite this, the power density output dramatically depends on the temperature. Measuring the temperature locally in an EESS will provide information contributing to the optimisation of the charging and discharging processes. To avoid disturbance in the flow through the flow cell system, ideally the temperature sensors have to be very small and robust. Optical fibre-based sensors offer a unique way to achieve this, as they are small, robust, accurate and inert to many harsh environments.

## Sensor Design

Three optical fibres were constructed with three fibre Bragg gratings (FBGs) on each fibre. These were calibrated for measuring absolute temperature and integrated into an RED/ED flow cell battery (Fig. 13.2).

Creating a periodic perturbation in the refraction index of the core (grating corrugations) of a single-mode optical fibre causes a successive, coherent scattering of a narrow band of the incident field [8]. The most robust reflection occurs at the Bragg wavelength, calculated by:

$$\lambda_B = 2\eta_{eff}\Lambda \tag{13.16}$$

Here, $\lambda_B$ is the Bragg wavelength, $\Lambda$ is the periodicity of the Bragg grating and $\eta_{eff}$ is the modal index of the optical fibre (Fig. 13.3). Therefore, the reflected wavelength is dependent on the modal index and the Bragg grating in the fibre.

These can both be influenced by temperature and strain; therefore, the most robust reflection (i.e. the Bragg wavelength) changes as a function of strain and temperature [9]. The change in Bragg wavelength as a function of strain and temperature can be determined by:

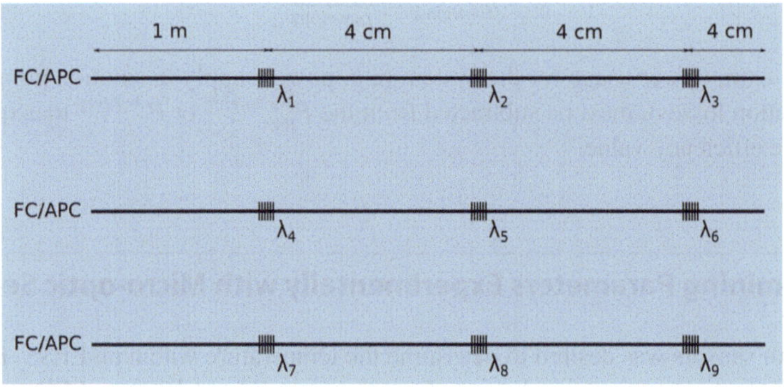


**Fig. 13.2** Schematic of the grating location on the fibres

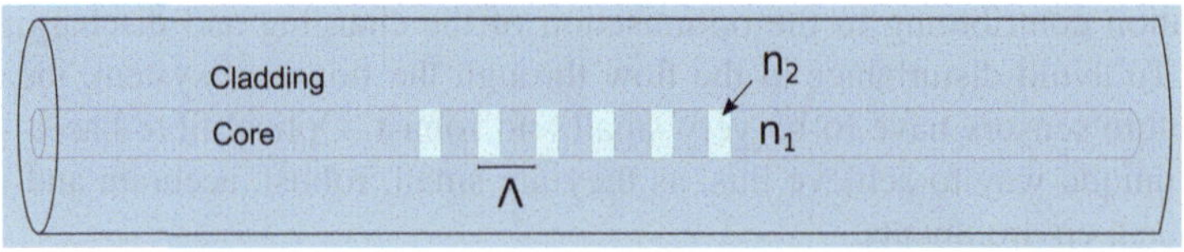


**Fig. 13.3** Schematic of an optical fibre with inscribed fibre Bragg grating

$$\Delta\lambda_B = 2\left(\Lambda\frac{d\eta_{eff}}{dT} + \eta_{eff}\frac{d\Lambda}{dT}\right)\Delta T + 2\left(\Lambda\frac{d\eta_{eff}}{dc} + \eta_{eff}\frac{d\Lambda}{dc}\right)\Delta c \quad (13.17)$$

The dependence of the Bragg wavelength is linear and does not have hysteresis.

## Experimental Setup

A light source emitting a signal with an 85 nm half bandwidth signal centred at 1550 nm is used for the sensor system. The light source is connected to the three Bragg grating fibres (Fig. 13.2), through a 125 μm single-model optical fibre. The signal is split twice by 50:50 couplers to obtain three separate signal inputs/outputs, sufficient to couple to the three Bragg grating fibres. The three fibres with Bragg gratings are placed inside an EESS between two membranes, so the Brag gratings are distributed evenly throughout the EESS cell. In addition to the light source, a spectrometer is connected before the first 50:50 coupler. Figure 13.4 shows the layout of this setup in detail.

The signal travels from the source, through the 50:50 couplers and along the three fibres with the Bragg gratings to the end of these fibres. At the end of each of the three fibres is a reflector. This reflects the signal through the fibres and couplers to the spectrometer. The spectrometer has a sensitivity range that includes the region of interest (1510–1595 nm; Fig. 13.5).

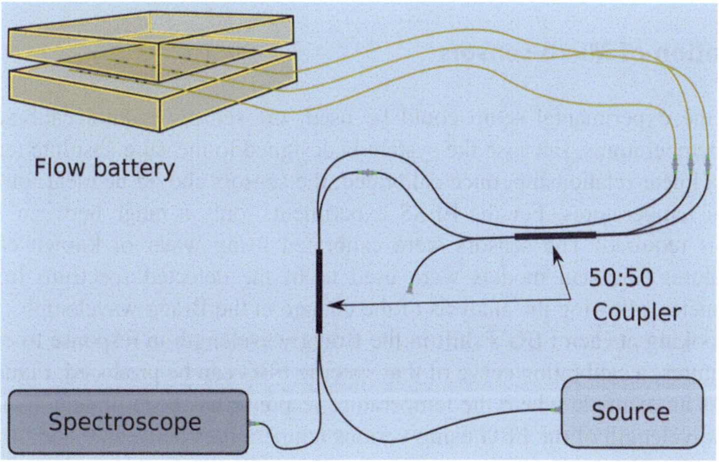


**Fig. 13.4** Schematic of the experimental setup

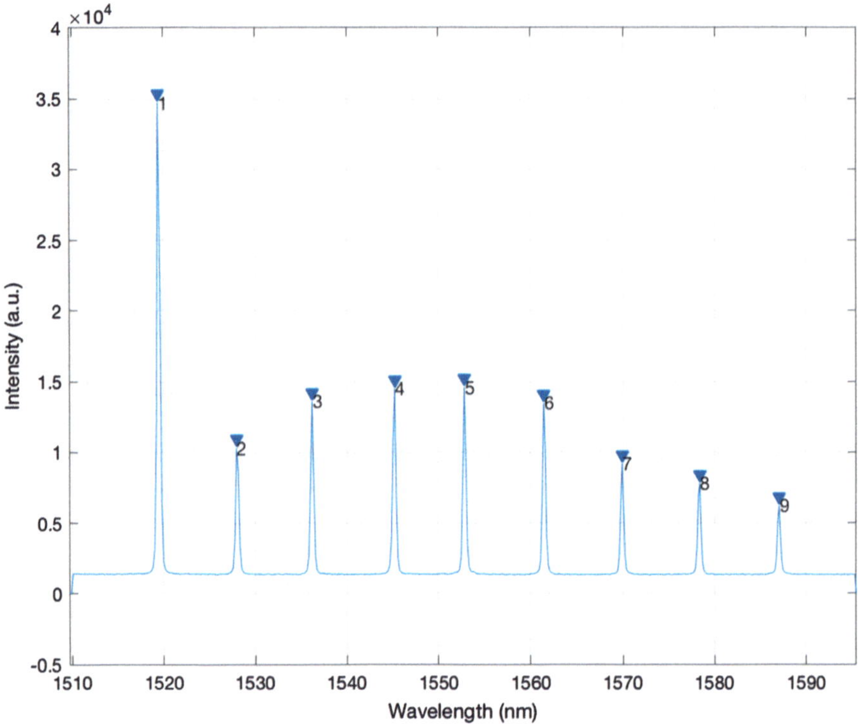


**Fig. 13.5** The reflected spectrum from the nine FBGs

## Calibration of Micro-sensors

Before the experimental setup could be used, the sensors require calibration to known temperatures. Because the system is designed to measure absolute temperature in a linear relationship, once calibrated, the sensors should be ideal for a wide range of temperatures. For the EESS experiments, only a range between 10 and 60 °C is required. The sensors were calibrated using water of known constant temperature. Gaussian models were used to fit the detected spectrum from the spectrometer, allowing the analysis of the change in the Bragg wavelength.

By looking at each FBG's shift in the Bragg wavelength in response to external temperatures, a calibration curve of that specific FBG can be produced. Figure 13.6 serves as an example where the temperature response has been plotted against the Bragg wavelength of the FBG using various temperatures between 0 and 60 °C.

Using this plot, temperature sensitivity can be determined as a function of change in the Bragg wavelength. Here, the sensitivity of the FBG is 0.01022 nm per °C, with a standard deviation of $7 \times 10^{-5}$ nm per °C. The root mean square can also be calculated from such linear regression plots. For this particular FBG, the root mean square was $3.1 \times 10^{-3}$ nm, which means that the FBG sensor has a precision of ± 0.3 °C.

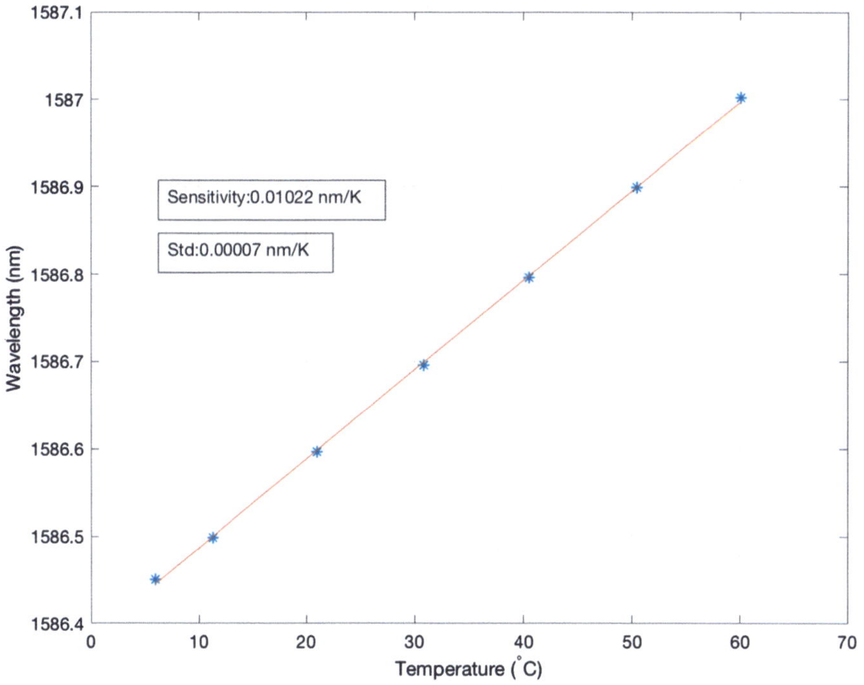


**Fig. 13.6** A plot of the peak Bragg wavelength as a function of temperature

## Implementation of Sensors into a Reverse Electrodialysis Cell

The network of nine FBG sensors was distributed evenly between two membranes and installed into the EESS flow cell, as shown by the schematic in Fig. 13.4. Room temperature solutions of NaCl at 5 M (for the concentrated flow cell) and 0.05 M (for the dilute flow cell) were filled into the EESS and left for 30 s for the temperature reading of the FBGs to stabilise. The FBG sensors were wedged between the concentrated flow cell and the dilute flow cell, separated only by the ion exchange membranes (one for the cation exchange and one for the anion exchange).

At the start of the experiment, all nine FBG sensors displayed the same stable temperature (room temperature). After 30 s, 80 °C solutions of NaCl at 5 M (for the concentrated flow cell) and 0.05 M (for the dilute flow cell) were slowly pumped continuously into the EESS. The temperature inside the EESS began to increase as the room temperature solutions were exchanged for the heated solutions. After 5 min, pumping of the heated solutions was stopped, and cooler solutions were pumped in until finally, all pumps were turned off, and the system relaxed to an elevated temperature relative to room temperature. The temperatures obtained from the nine FBGs are shown in Fig. 13.7, where the plot shows the change in temperature relative to the room temperature (indicated as 0 °C).

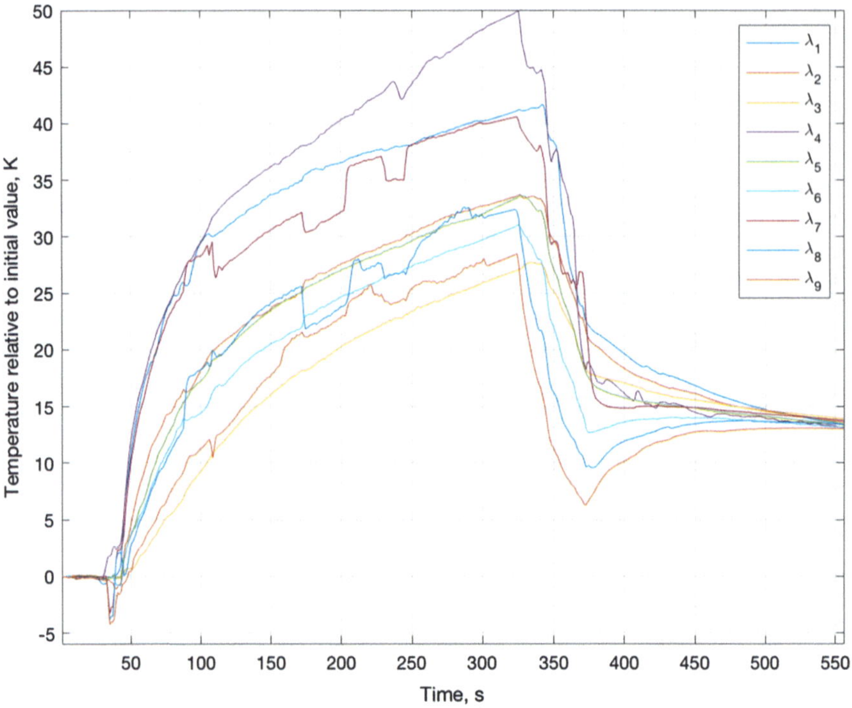


**Fig. 13.7** The temperature profile of nine FBGs inside an EESS flow cell battery

The plots shown in Fig. 13.7 are indicative of the distribution of temperature within an EESS when solutions of different temperatures flow through the cell. As can be observed here, there were significant discrepancies between different FBG sensors, yet all sensors started and ended at the same temperature. These discrepancies could be the result of the drastic distribution in temperature within the cell, or they could be due to the turbulent flow of the solutions. The turbulent flow could cause the fibres to flex, causing strain at the FBG sensors. As stated earlier, the Bragg wavelength is a function of both temperature and strain; therefore, these measurements may not be entirely due to temperature changes.

## Conclusions

Due to the temperature in the EESS affecting the performance of the flow cell, a profile and distribution of temperatures inside the system are required to improve computational modelling methods for optimising EESSs. Here, micro-optical sensors have been suggested to be able to fulfil this requirement due to their small geometrical size, their inertness and their accuracy. A preliminary study was performed and showed FBG sensors that have been calibrated for sensing temperature could be implemented into an EESS. The temperatures from each FBG were recorded in real time, and an observable temperature gradient was displayed. Despite this, there is potential that the temperature profiles had reduced accuracy due to the potential strain caused by the flow of solutions in the cell. The turbulent flow could cause the FBGs to flex, as they have limited structural support in the EESS used. By implementing further structural supports for the FBG sensors to avoid strain caused by the flow of solutions, it is believed that the performance of FBG sensors will significantly improve inside the EESS.

**Acknowledgements** The authors are grateful to the ENERSENSE programme and NTNU Team Hydrogen at the Norwegian University of Science and Technology (NTNU) for supporting and helping on this book project.

## References

1. Kingsbury RS, Chu K, Coronell O (2015) Energy storage by reversible electrodialysis: the concentration battery. J Memb Sci 495:502–516
2. van Egmond WJ, Starke UK, Saakes M, Buisman CJN, Hamelers HVM (2017) Energy efficiency of a concentration gradient flow battery at elevated temperatures. J Power Sources 340:71–79
3. Kamcev J, Paul DR, Manning GS, Freeman BD (2018) Ion diffusion coefficients in ion exchange membranes: significance of counterion condensation. Macromolecules 51(15):5519–5529
4. Giorno L, Drioli E, Strathmann H (2016) Water transport in ion-exchange membranes. In: Drioli E, Giorno L (eds) Encyclopedia of membranes. Springer, Berlin
5. Pitzer KS, Pabalan RT (1986) Thermodynamics of NaCl in steam. Geochim Cosmochim Acta 50 (7):1445–1454
6. Ramon GZ, Feinberg BJ, Hoek EMV (2011) Membrane-based production of salinity-gradient power. Energy Environ Sci 4(11):4423–4434
7. Post JW, Hamelers HVM, Buisman CJN (2008) Energy recovery from controlled mixing salt and fresh water with a reverse electrodialysis system. Environ Sci Technol 42(15):5785–5790
8. Hill KO, Meltz G (1997) Fiber Bragg grating technology fundamentals and overview. J Light Technol 15(8):1263–1276
9. Kinet D, Mégret P, Goossen K, Qiu L, Heider D, Caucheteur C (2014) Fiber Bragg grating sensors toward structural health monitoring in composite materials: challenges and solutions. Sensors 14(4):7394–7419

# Target Audience

The general target audience for this Primer is industry professionals, researchers and graduate students, especially those starting a new topic of research or coming into the field.



J. J. Lamb, B. G. Pollet (eds.), *Micro-Optics and Energy*,
https://doi.org/10.1007/978-3-030-43676-6

The manufacturer's authorised representative in the EU is Springer Nature Customer Service Centre GmbH, Europaplatz 3, 69115 Heidelberg, Germany. If you have any concerns regarding our products, please contact ProductSafety@springernature.com

Printed and bound by CPI Group (UK) Ltd, Croydon, CR0 4YY
05/10/2026
02228124-0004